FMD p95

# AUDIO TECHNICIAN'S BENCH MANUAL

# AUDIO TECHNICIAN'S BENCH MANUAL

### JOHN EARL

FOUNTAIN PRESS:LONDON

Fountain Press
46/47 Chancery Lane
London WC2A 1JU

First Published 1972

© John Earl 1972

ISBN 0 852 42093 5

Printed in England by Offset-Litho at
Page Brothers (Norwich) Ltd.

# CONTENTS

PREFACE

1 EARS AND TEST INSTRUMENTS  1
Subjective choice. Loudspeaker testing. Harmonic distortion. Class B amplifier. Transient type tests. Filters. Instruments for amplifier tests. Instruments for tuner tests. Instruments for other tests.

2 AMPLIFIER TESTS  33
Basic functions. Output tests. Instability. Dummy loads. Power amplifier tests. Voltage regulators. THD testing. IMD testing. Power bandwidth tests. Frequency response measurements. Damping factor. Sensitivity voltage test. Maximum input voltage tests. Hum and noise. Square-wave testing.

3 TUNER TESTS  72
F.M. tuner tests. Audio signal filtering. Frequency response. Stereo subchannel rejection. Signal-to-noise ratio test. Generator matching. Sensitivity tests. Capture ratio. Selectivity. Spurious responses. Automatic frequency control. F.M. stereo tests. F.M. stereo S/N ratio. A.M. sections. Tuner-amplifiers.

4 DISC PLAYING EQUIPMENT TESTS  108
Pickup tests. Trackability. Tracking weight. Stylus tip impedance. Side-thrust correction. Compliance tests.

Effective tip mass estimating. Frequency response tests. Stereo separation tests. Output voltage tests. Tone-burst and squarewave tests. Pickup arm tests. Turntable unit tests.

5 SYSTEM TESTS 137

Tuner amplifier tests. Hum and noise. Radio breakthrough. System faults. Impulsive interference. Channel balance tests. Stereo identicality factor. System crosstalk tests. Frequency response tests. System distortion and hum level tests. Acoustic feedback tests. Impedance tests. Tone controls and filters. Tape recorder tests. Acoustical characteristics.

6 AUDIO STANDARDS AND DEFINITIONS 166

DIN 45–500. General requirements. F.M. tuners. Record players. Tape recorders. Microphones. Amplifiers. Loudspeakers. Performance tests for complete systems.

INDEX 179

# PREFACE

WHEN WRITING THIS BOOK my plan was deliberately to concentrate more towards the testing of audio equipment and example test procedures than to expound the general principles of equipment design and servicing. The practising audio technician is already a proven master of his craft. For students and enthusiasts desirous of extending their knowledge in this direction there exists a number of excellent books written specifically with them in mind.

However, there is a significant gap in the literature between this sort of book and that dealing with advanced design techniques, particularly with regard to the detailed testing of audio equipment, and it is towards filling that gap that this book is aimed.

I have devoted the first chapter to subjective appraisal and test instruments, while following chapters highlight hosts of tests which can be applied to all the departments of an audio system. Some of the tests are already commonly used and some are based on procedures recommended by various 'standards' authorities, but interpreted and expanded to reveal suggested and proven instrument setups. Others are tests which the 'standards' authorities have yet failed to embrace and which I have myself devised over the years as a means of securing a complete objective appraisal of the audio art as a whole.

Where appropriate I have compared the procedures of parameter tests suggested by two of the main 'standards' authorities, namely, the British Standards Institution and the Institute of High Fidelity Manufacturers Inc. (IHF, USA), and in certain cases have discussed probable causes of sub-standard performance and adjustments for improvement.

In addition to 'audio electronics', I have included a large chapter on testing the mechanical and electro-mechanical items of disc playing equipment. The final two chapters deal with tests in complete systems and audio standards generally.

The plan, then, has been to create a book for audio technicians which, it is hoped, will find itself for most of the time on the bench. In short, a 'bench manual'. Obviously, a good deal of the information will be of assistance to the audio equipment user and hi-fi enthusiast, enabling them and the hi-fi dealer at large to gain an insight into the various parameters which are so extensively specified by the equipment manufacturers but not always adequately understood.

Finally, I would like here to record my sincere thanks to the instrument and equipment manufacturers and to my friends in the industry for their help and encouragement, without which this book would never have left the ground.

John Earl                                                           *1972*

CHAPTER ONE
# EARS AND TEST INSTRUMENTS

IT IS HELD BY SOME PEOPLE that the most meaningful tests of audio and hi-fi equipment are those which engage the ears of the listeners. It is often claimed that ears can discern the subtle characteristics of the reproduction which even the most sensitive instruments are incapable of detecting. It is perfectly true, of course, that our ears are the final arbiters in the complex reproducing chain. Sound, like vision, smell and taste, is totally subjective. That is, it is a personal experience.

**Subjective Choice**

Since the advent of colour television more and more people have come to realise just how subjective colour vision is. This medium is demonstrating how differently we see or appreciate colours, and is a good subjective example. Colour receivers are equipped with controls which make it possible to alter the colour saturation, the background 'white' and sometimes the hue, and based on a standard transmission it is surprising the different settings given to these controls by different operators for a colour display that each one feels is correct.

Instruments are available for establishing the correct 'white', but only the eyes of the viewers can determine when the parameters of the actual colours are right—as they see them.

One large chemical plant in an industrial area endeavoured by the use of special instruments to investigate complaints from local residents of offensive odours. Because the sense of smell is so subjectively based this scheme failed to work, and ultimately the plant was obliged to employ a team of trained operators to locate and analyse the smells subjectively. Similarly, there is only one way to match the vintage and blend of wines—subjectively, by the sense of taste.

In all these instances a comparison is required for the optimum appraisal—the original colour test card for colour television, the original chemicals for the offensive odours and for sound reproduction the original music as heard in the concert hall.

In order meaningfully to appraise the quality of reproduction of an audio system, therefore, one must have in mind just how the original performance sounded. The audio technician should thus have an ear for music, and this applies in the greatest force when he is dealing with the most subjective item of all—the loudspeaker system.

## Loudspeaker Testing

It is possible to apply instruments for loudspeaker testing, leading to sound pressure and frequency response curves over the audio spectrum, on axis, at various points off axis over 360 degrees and in various environments from the normal listening room to an anechoic room, which is one without reverberation. The curves resulting from such tests might well have scientific value, but they tell the user very little about how the loudspeaker will sound in his own lounge when working from a particular amplifier and programme signals.

Various objective tests have been devised for loudspeaker systems, including interrupted tone tests, distortion tests, etc., but of recent years loudspeaker manufacturers have come to realise that while such tests are often necessary for the development of a new system they cannot be used as a guaranteed prediction of how the loudspeaker will sound under normal listening conditions.

## Comparative Tests

The design trend is now to integrate the objective tests with more subjective ones, and a recent scheme is based on comparing the sound produced by a loudspeaker system under development with the real sound. Some readers may recall that this was a pet exercise of Gilbert Briggs, one of the loudspeaker pioneers of Wharfedale fame, during some of his loudspeaker demonstrations at the start of the Festival Hall. He often had his audience guessing as to whether they were listening to the real thing or to the reproduction through one of his latest loudspeaker system creations.

Testing of this kind has become more sophisticated, although basically unchanged in technique, and now the original sound may be recorded, with the loudspeaker under test reproducing the original recorded sound.

At the time of writing controversy is keen on the subject of

## Ears and Test Instruments

directional and omnidirectional loudspeaker systems, particularly relative to the omni-type which rely on reflected sounds for their true function and stereo reproduction. There is no doubt that some listeners prefer the greater 'spread' of sound provided by the omnis at the expense of a widened and impaired stereo image, even though it can be argued that only those systems which fail to rely so much on reflected sounds are scientifically more suitable for the reproduction of stereo signals. The illustration here is not itself intended to be controversial, but is included merely to highlight that it is how the system sounds to the listener that matters most, in spite of the function possibly being scientifically inaccurate.

It must be mentioned, however, that there are some systems which are not totally omni-directional and which do not depend so much as the true omni on reflected sounds, though not being so directional as the basic loudspeaker system. With these the sound front is relatively wide, and the design is to simulate a natural sound source, so from the stereo aspect they are less scientifically in error than those designs relying essentially on reflected sounds.

### Extension of Senses

Loudspeaker system subjectivity is emphasised when it comes to choice, but as this subject has been covered in my companion volume *Pickups and Loudspeakers* there would be little point in pursuing it here.

Obviously, there are hosts of functions in audio equipment which are not directly available to the senses, whether right or wrong. Moreover, as it is an audio technician's task to locate and remedy fault conditions, the senses need to be extended by instruments, for such can be regarded as contrivances for translating into a form acceptable by the senses functions of the circuit and system which are not.

Instruments, of course, are also required for testing and appraising the various items of an audio system to make sure that repair or adjustment is, in fact, necessary, and that each item fully meets its specification after a repair or adjustment has been performed.

### Harmonic Distortion

When one is dealing with equipment of specifically high quality it is not always obvious by listening tests alone that the equipment is working according to its specification. For example, the specification may give the total harmonic distortion (THD) as a mere $0.03\%$ at 1kHz, while due to a minor circuit variation it might have risen to 0·2%. Unless one is very conversant with the amplifier under its normal conditions of operation it would be virtually impossible to say conclusively that the THD is higher than the specification merely by listening tests.

On the other hand, the critical listener whose amplifier it is would be more aware of the error of the reproduction. This is because he has been accustomed to the reproduction before the fault occurred—hence his need to consult the technician!

In spite of this subjective awareness by the owner the technician must also be equally convinced that the amplifier is below par before he can do anything about it, for he will tell you that it is not uncommon for a hi-fi amplifier shortcoming to be blamed for deteriorated sound quality, when in actual fact a worn pickup stylus has been responsible.

To prove that the amplifier is really in trouble, therefore, the technician must have available test instruments capable of measuring accurately the items of the specification. Without this sort of equipment he might well be able to repair an amplifier—that is, to correct severe distortion, bad noise or complete failure—but he will be totally in the dark with regard to the *smallness* of the distortion and noise, etc.

**Critical Approach**

Today's audio technician is concerned mostly with high quality equipment of the hi-fi calibre and thus differs significantly from his colleague in the television or radio workshop where sound distortion of several per cent total harmonic and more intermodulation is daily tolerated without grimace—and accepted as normal. The audio technician, to whom this book is directed, is a much more critical animal (from the sound point of view, anyway). In order to handle the job properly he must himself be thoroughly interested in hi-fi reproduction in all its aspects.

The chaps in the television and radio workshop not uncommonly regard us as being a bit over the top. They cannot understand why it is that we spend hours in reducing distortion, say, from 0·1% to 0·02% when the distortion on the programme signals is possibly several magnitudes above this low level. But we know very well that the hi-fi craft is for ever striving to make the reproduction as close as possible to the original, and that this means keeping the distortion in *all* the departments at the lowest possible level, amongst other things.

Sadly, instruments tend to work the opposite way to our critical sense of hearing, for it is not uncommon for a distortion factor meter to indicate a relatively low THD yield and yet for the reproduction to sound as if it is carrying distortion of a far higher percentage. This is because there are some distortion waveforms which occur too quickly for the meter to respond yet which are distinctly detected by our ears. The implication that there might not be an absolute correlation between the sounding of an audio system and what the meter tells is true; again, making the ear the final arbiter!

Moreover, the nature of the load presented to an amplifier when instrument tests are conducted may differ from that of the load presented by the loudspeaker under normal operating conditions. The nature of the test signal is also different from that of music waveforms, and since we can rarely employ music signals for our tests we can never really discover the genuine nature of the reproduction by instrument tests. Nevertheless, we can get close to a meaningful assessment by using special test signals and loads simulating the loudspeakers likely to be connected to the amplifier.

## The Class B Amplifier

Obvious lack of objective/subjective correlation became increasingly apparent with the advent of the Class B solidstate amplifier. Class B implies that the push-pull transistors are biased approximately to collector current cut-off under zero input signal condition, and that the collector current in one device of the pair flows approximately for one half of each cycle, while the partnering device is driven further into cut-off, the conducting mode of the pair then reversing on the other half of each cycle.

This basic mode of working yields high odd harmonic distortion, with the distortion frequently having a maximum value at low power output. The reason for this stems from the discontinuity of the overall transfer characteristic in the middle region, and the shortcoming can be avoided by biasing the output pair so that a relatively small collector current flows in the absence of drive.

Contemporary practice automatically assumes that a Class B amplifier is one which is deliberately biased for a small quiescent current. Just how much quiescent current is necessary to endow the transfer characteristic with optimum linearity depends very much on the design of the amplifier, some requiring almost that value equivalent to the requirements for operation towards Class AB; that is, where the biasing and drive signal are such that collector current flows in the conducting device for more than a half, but less than the entire, input cycle. On the other hand, some better designs produce very low distortion even when the quiescent current is a small fraction of the full-drive current.

Class A operation, of course, implies that the biasing and drive signal cause the collector current of the transistor or transistors to flow at all times. Class B operation has the advantage of a high collector circuit efficiency—maximum 78% compared with 50% for Class A amplifiers.

**Quiescent Current Preset**

To keep the efficiency of Class B amplifiers as high as possible the quiescent current should be adjusted for the smallest value consistent with the least distortion over the dynamic power range. There is usually no particular reason why the quiescent current should be set above this value, so the preset control provided by many amplifier designers for this purpose is best adjusted while testing for THD and viewing the residual on an oscilloscope.

I shall be having more to say about this in Chapter 2, but right now I wish to highlight the theme earlier expounded in that an amplifier may have a low THD read-out yet be responsible for reproduction below hi-fi standard.

Fig. 1.1 shows at (a) and (b) THD residual of somewhat sharp waveform, while (c) shows a waveform carrying a distinct short-duration peak. The nature of these waveforms implies a high degree of high-order odd harmonic distortion, and most of them were obtained from Class B amplifiers either of not too good design or with the quiescent current below (or above) that required for the least distortion.

The THD residual waveform at (d) was obtained from a Class A amplifier. This is devoid of sharply rising sides or peaks, which implies that the distortion is of lower-order harmonic, being composed mainly of second and third harmonic.

Now, owing to the method of measuring THD it is possible that the waveform at (c), for instance, will give a smaller reading than that at (d), but there is little doubt that the amplifier responsible for waveform (d) will be less prone to stridence than the amplifiers responsible for waveforms (a), (b) and (c), though it needs a trained ear to detect the difference in less extreme cases.

This is certainly not to imply that Class A amplifiers are essential for top quality reproduction. Designers have now devised ways and means of completely taming Class B distortion which formerly arose from the middle of the transfer characteristic discontinuity or non-linearity.

In fact, when analysing the distortion from some recent Class B designs of high quality I have found it virtually impossible to detect significant differences in the make-up of this from that

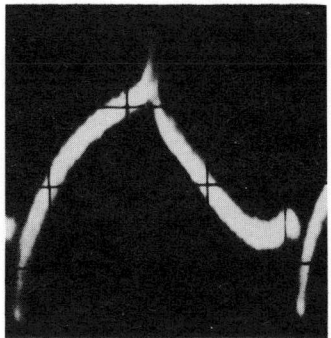

Fig. 1.1 THD residual. (a) and (b) indicate high-order odd harmonics. (c) shows short-duration pulse due to shortcomings in Class B amplifier. (d) is waveform obtained from a Class A amplifier, composed mostly of second and third harmonics.

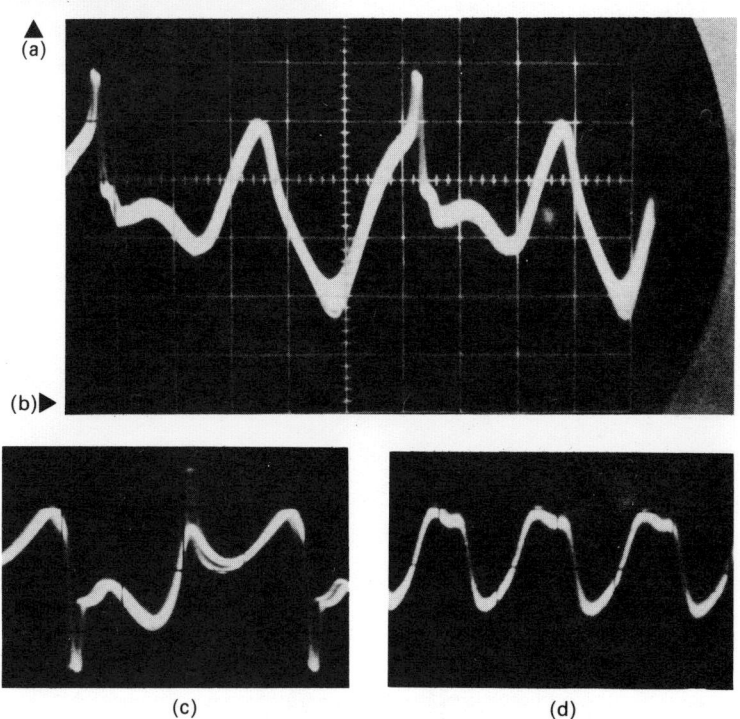

produced by Class A amplifiers.

A well designed Class B hi-fi amplifier should have mostly second and third harmonic yield, with very little high-order odd harmonic components when the quiescent current is set correctly. However, the distortion can rise above the minimum possible if the quiescent preset is maladjusted, which is a point well worth bearing in mind.

## Weighting

It has been suggested that a weighting factor should be applied to THD measurements, so that the read-out is more accurately geared to the subjective annoyance value, but so far as I know this is rarely—if ever—adopted in general testing as it requires measurement of harmonics up to about the tenth order—a time consuming task! (Readout proportional to rate of change of the THD has been suggested. This would appear to represent a better weighting.)

A weighting factor is sometimes applied to noise measurements however. Hum and noise below the rated output of the amplifier is commonly measured by a very sensitive millivoltmeter and expressed as a dB ratio. A ratio of 60dB, therefore, implies that the summed voltages of the hum and noise across the output load represent a value 1,000 times less than the voltage across the load when the amplifier is fully driven; or that the power of the hum and noise is 1,000,000 times less than the full power of the amplifier.

Unweighted, the hum and noise are measured over the full bandwidth of the amplifier, but when weighted the output power or voltage is measured with the read-out device preceded by, or incorporating, a frequency weighting network with a response

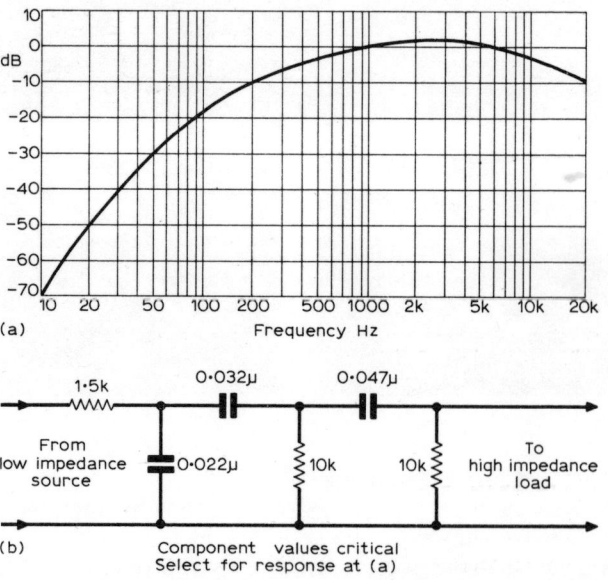

Fig. 1.2 Weighting curve (a) and circuit (b).

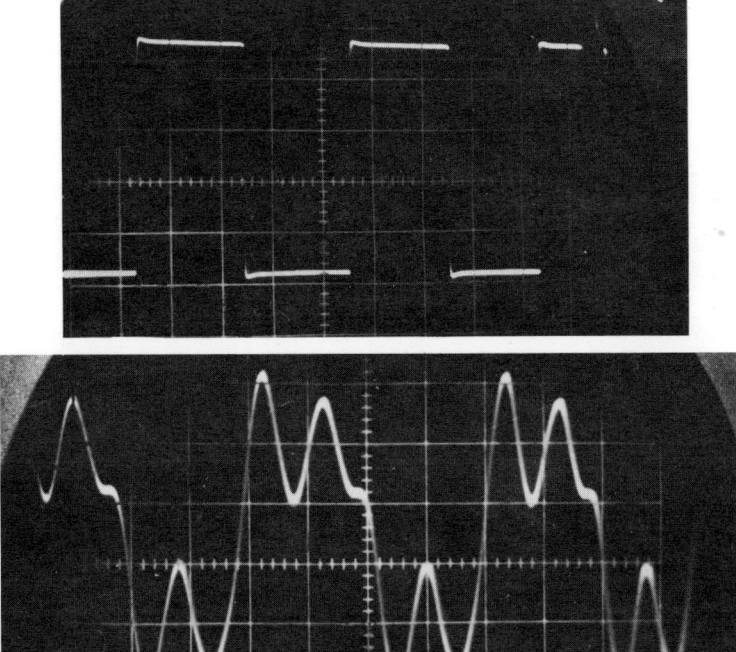

Fig. 1.3 Squarewave oscillograms. (a)—above—input waveform. (b)—below—high amplitude ring across reactive load.

characteristic similar to that shown by the curve in Fig. 1.2.

This passes the noise signals in the part of the spectrum over which human hearing is the most sensitive, while heavily attenuating the lower frequencies which have smaller subjective annoyance value. Clearly, then, a given amplifier will have a higher weighted ratio than an unweighted one.

**Transient-type Tests**
As already mentioned, some amplifiers tend to overshoot or ring badly on reactive loads when fed with transient-type signal. To test for this a pulse-type or squarewave signal needs to be applied and the output monitored on an oscilloscope across a load comprising reactance to simulate that of a loudspeaker system.

Fig. 1.3 shows at (a) an input squarewave and at (b) the waveform across a reactive load at the output, where there is almost two cycles of ring. Some amplifiers ring even more severely than this, and some less, as shown by the waveforms (c) and (d). Good amplifiers should not overshoot or ring significantly when looking into purely resistive loads, but a small percentage overshoot (less than one complete cycle) not uncommonly occurs on the best of amplifiers when the load is heavily reactive. The percentage overshoot or ring is the ratio of the amplitude of the squarewave to the amplitude of the overshoot or ring. The larger the ratio, obviously the better.

Some amplifiers are critical to load reactance, and while they may not ring excessively on heavy reactance, severe ringing might be encouraged by relatively small values of capacitance and/or inductance, with resistance, and if these are approximated by the loudspeaker with its crossover network the tonal quality of the reproduction could be affected.

The rings constitute damped oscillations, and since they are spurious signals they are bound to affect the reproduction in some way or other.

It is noteworthy that amplifiers with wide power bandwidth, provided by transistors of high $f_T$ (gain-bandwidth product) and plenty of negative feedback, require more detailed design for small percentage overshoot or ring than amplifiers whose bandwidth is restricted by the nature of the transistors or deliberately by the use of a slow roll-off low-pass filter.

Thus, if a relatively inexpensive amplifier appears by squarewave tests to have a remarkably good transient performance check on its power bandwidth before concluding that a more expensive design has a poorer standard of performance!

### Need for Extended Response

While our hearing goes little beyond about 16kHz (it diminishes with frequency as we get older), the power response of the amplifier should significantly exceed that frequency in order to preserve the small rise and falls times of transient-type music signals.

Such signals can be analysed as a multiplicity of sinewaves extending from the fundamental in harmonic progression to quite high orders, and to preserve the shape of the compounded signal all the harmonics must be passed by the amplifier without their amplitude or phase relationships being unduly modified.

If there is modification the compounded signal will have a waveform different from that of the original signal, so the character of the sound might alter, depending on the extent of the modification.

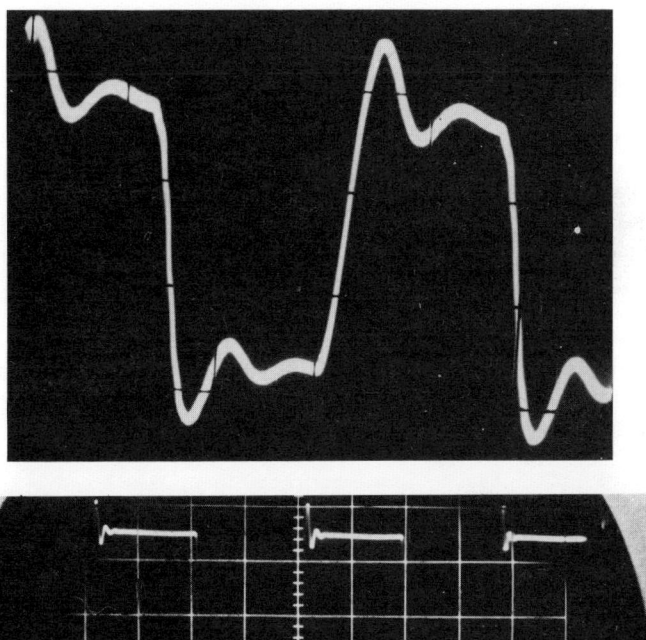
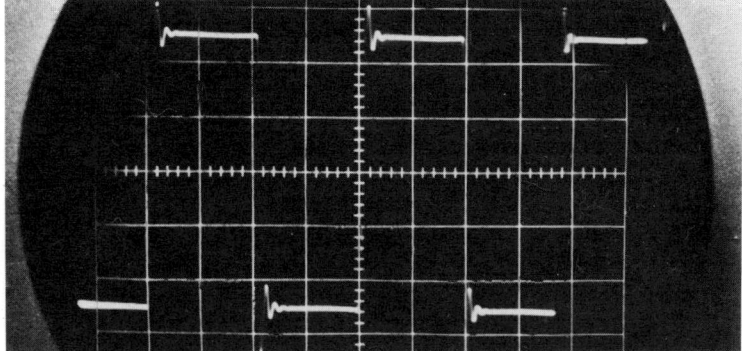

Fig. 1.3 (c) and (d) less severe rings across reactive loads than those shown in Fig. 1.3 (a) and (b).

Since the staccato characteristics of music are contained within rapidly rising and diminishing waveform components, an amplifier lacking sufficiently high-frequency power response sounds 'dull' and is without 'attack'.

### Filters

Nevertheless, it is often necessary to roll-off the top treble by a low-pass filter to remove excessive distortion or noise present on the programme signal. To avoid the wanted mid-treble signals from being too severely attenuated the filtering should not come on gradually and then slowly roll-off into the high-treble.

On the other hand, if the roll-off is too sudden and unduly fast the resulting phase shift in the circuit can encourage rings similar to those which I have been discussing. Again, these can change the tonal quality of the reproduction.

Some of the best filters are engineered in steps, starting at about 6dB/octave and then becoming more rapid, reaching an ultimate rate of 12 or 18dB per octave at the top treble end of the spectrum. This nature of design makes the filter less ring-prone; but in any case, a low-pass filter should only be used when the filtering gives a better sound than the signal passing through the whole bandwidth of the amplifier; that is, when the programme signal carries a lot of noise or distortion.

The high-pass filter—that which rolls-off the bass end of the spectrum for eliminating sub-bass noises such as turntable noises and rumble—can also exhibit phase shift effects, but filtering at this end of the spectrum is less critical in this respect. Nevertheless, the roll-off should not commence too early, for then it would attenuate the wanted bass signals above the rumble and low-frequency signals that its function is to remove.

The turnover frequency should be around 30Hz and the rate of roll-off about 12dB/octave, or higher. Some amplifiers have the high-pass filter included only in the magnetic pickup preamplifier circuit, and then often unswitched.

**Necessary Instruments**

We have seen that our ears play a major role in the determination of the overall performance of an audio system, but the audio technician must obviously possess instruments both for the measurement of the various items of the specification and for fault finding and adjustments.

Although some of the instruments necessary for the basic service procedures will be suitable for the measurement of the performance, a better class of instrument—more of a laboratory quality—is becoming available from various sources for the latter. This means, therefore, that while some of the instruments already in use, perhaps, for radio and television servicing, will have some purpose for audio work, additional instruments will be required for detailed analysis. Put another way, we can be pretty sure that when we have the instruments needed for detailed measurements, then we shall be well equipped for fault finding and adjustments.

The plan now is to look briefly at some parameters of various items of audio equipment to discover what instruments and ancillary items are required for their measurement.

## INSTRUMENTS FOR AMPLIFIER TESTS
### Power Output Tests (r.m.s.-based)

Instruments required for these tests are a sinewave generator for producing a test signal, a dummy load for the amplifier and an audio voltmeter reading up to at least 20V r.m.s. (alternatively a wattmeter suitable for audio signals carrying its own switchable load can be used. When a dummy load and audio voltmeter are used the audio power in the load is calculated in terms of $V^2/R$, where $V$ is the r.m.s. voltage across the load and $R$ the resistance of the load in ohms). An oscilloscope for observing when the amplifier reaches sinewave clipping power is another requisite.

It is sometimes necessary to check the power output of a stereo amplifier when both channels are being driven to capacity together. In this case two dummy loads are required (one for each channel). It is normally possible to use the single sinewave generator to drive both channels simultaneously, and the single audio voltmeter (or wattmeter) can then be switched from one channel to the other to obtain the separate readings.

Note that the rated power output of an amplifier may be specified with both or only one channel driven. Moreover, the continuous power output may be less than the maximum power output. The latter is commonly taken as the power yield prior to the onset of waveform clipping or when the THD is at a specified value. The r.m.s.-based power is sometimes referred to as *steady-state power* or *continuous power*.

### Power Output (Music Power)

It is held by some people that r.m.s.-based power is not a true indication of the performance of the amplifier under music signal conditions, since it gives undue emphasis to the ability of the amplifier to heat a load with a signal of low distortion.

Various schemes have thus been devised to measure the power output under controlled signal conditions simulating the nature of a signal produced by music. We thus not uncommonly find in an amplifier specification a 'music power' rating instead of, or in addition to, the continuous power rating.

The most frequently used is that by the American *Institute of High Fidelity, Inc.,* abbreviated IHF. Instruments required for testing to this specification include those already mentioned for testing continuous power, but with the addition of a special modulator for producing the music-type signals from a sinewave source and a distortion meter, since one aspect of the measurement is based on transient-distortion testing. The modulator is also used to trigger a single horizontal sweep of the oscilloscope. The IHF rating, therefore, can be regarded as 'short duration output power'.

Conversely, this scheme is not favoured by some people since it tends to give optimistic results and fails to take into account the loading of the supply rail of the amplifier during sustained loud passages of music.

One alternative is the use of interrupted sinewave signal with a duty period of 22%, meaning that the sinewave is on for 22% of the time and off for 78% of the time. This method of testing is used by Alba (Radio and Television) Ltd., but, of course, requires the extra use of a pulse generator of some kind.

Music power is sometimes referred to as *dynamic output*, but the test most favoured to date is that of continuous power based on the r.m.s. voltage across the output load. Roughly, an amplifier with a continuous power rating of, say, 20W would have an IHF Music Power rating around 30W.

**Power Bandwith**

Instruments required are a sinewave source variable in frequency over the audio spectrum. Output load or loads of suitable power rating (see Fig. 1.4). Audio voltmeter (maximum reading at least to 20V r.m.s. allowing a maximum of 50W to be measured into 8 ohms) preferably with a dB scale. Distortion measuring set if the bandwidth limits are to be defined relative to a given level of distortion. Oscilloscope to check for waveform clipping level.

**Damping Factor**

Sinewave input source (adjustable to 50Hz when the measurement is to be made according to the British Standard 3860:1965). Output load which can easily be switched in and out of circuit and which corresponds to the rated load resistance of the amplifier (nominally 8 ohms with transistor amplifiers) and of a wattage to handle one-quarter of the rated power output. Audio voltmeter with a full-scale range near to 4V r.m.s.

**Source Resistance (or Output Resistance/Impedance)**

The source resistance of an amplifier is equal to the rated load resistance divided by the damping factor, so when the damping factor has been measured as above, the source resistance can be calculated based on the value of the output load resistance used. For example, a damping factor of 80 referred to an 8Ω load corresponds to a source or output resistance of 0·1Ω.

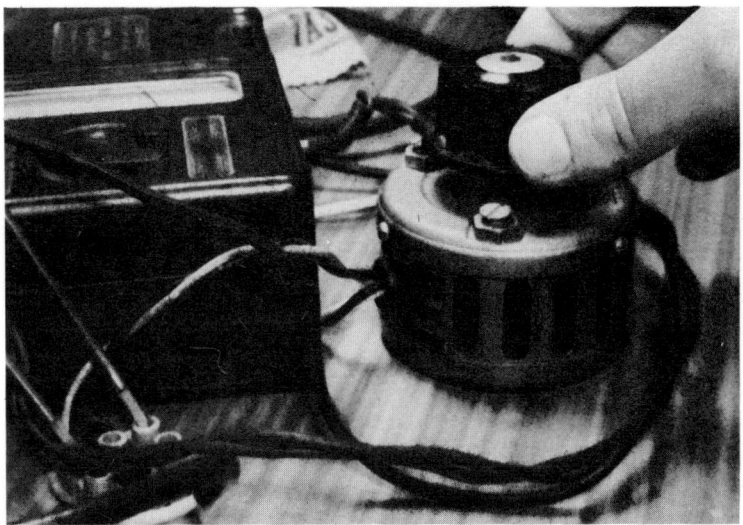

Fig. 1.4 A high-wattage variable resistor (rheostat) like this is useful for amplifier loading when a suitable instrument is available to set the required resistance value accurately.

The output *impedance* is sometimes measured at 1kHz by means of an audio-frequency bridge when the voltage across the output terminals of the amplifier does not exceed that corresponding to the rated output power. The value is sometimes dependent on the amplitude of the test signal.

**Total Harmonic Distortion**

The basic instrument required for THD tests is the distortion factor meter or test set. Most instruments of this kind require a sinewave signal generator of very low distortion to work with them. BS3860:1965 states that the r.m.s. total of all components other than the fundamental should be less than one-fifth of the harmonic distortion of the amplifier to be tested. If a sinewave generator fails to meet this requirement it is possible to use an external filter to attenuate the unwanted harmonics.

The distortion factor meter removes the fundamental, leaving all the harmonics, the power in which is then summed and measured as a percentage of the output voltage.

Some THD meters are equipped with a readout device, while others need to be used in conjunction with an audio voltmeter (see

Figs. 1.5, 1.6 and 1.7). With instruments like the Sugden Si452, therefore, a very sensitive millivoltmeter, reading down at least to 100μV f.s.d. (there is one matching the Sugden Distortion Measuring Unit, see Fig. 1.8), is also required actually to measure the THD yield in terms of a dB ratio or percentage.

It is also necessary to measure the output power at which the distortion is being measured, so in addition to the instruments so far mentioned, we require a suitable output load and an audio voltmeter or wattmeter. It is sometimes possible to use the readout device of the distortion test set to adjust the output power prior to the distortion measurement, When an instrument like the Sugden Si452 is employed, the millivoltmeter used for THD readout can double as the audio voltmeter for power output measurement provided it has a sufficiently high-reading range.

It is most important to remember that the THD readout includes noise signals of the amplifier and test equipment. Thus the real THD may be less than indicated. This is because the 'notch' of the distortion factor meter merely removes the fundamental within a very narrow bandwidth, thereby leaving the amplifier and test setup wide open to noise over virtually the whole passband.

Facilities are generally provided so that the THD/noise residual can be monitored on an oscilloscope. This is very useful since it not only allows us to see how much noise is contributing to the

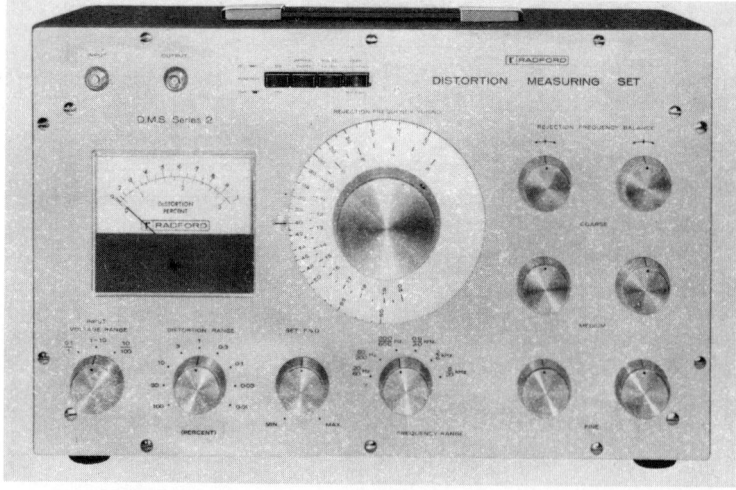

Fig. 1.5 Radford Distortion Measuring Set. This can read THD directly down to 0·01% f.s.d. It is battery powered.

readout, but it also makes it possible to analyse the nature of the THD and to determine the orders of the harmonics.

Moreover, it shows clearly when high-order harmonics or 'switching pulses' are being generated by a Class B amplifier in the non-linear transfer crossover region (see the waveforms in Fig. 1.1).

**Harmonic Distortion (Wave Analyser Method)**

A THD readout provided by a distortion factor meter is a very convenient method of measurement because it automatically sums the power in all the harmonics, and used in conjunction with an oscilloscope provides meaningful objective information on amplifier non-linearity and overall distortion performance.

The alternative is the wave analyser. This also has a tunable narrow-band filter, but instead of notching out the fundamental, as in the distortion factor meter, it is tuned to select one by one the individual harmonics. These can be expressed separately as a percentage of the fundamental or the total may be calculated to give the THD value.

THD is expressed as

$$\% \text{ THD} = \frac{\sqrt{(E_2^2 + E_3^2 + E_4^2 \ldots)}}{\sqrt{(E_1^2 + E_2^2 + E_3^2 + E_4^2 \ldots)}} \times 100$$

where $E_1, E_2, E_3, E_4$, etc., are the voltage (could be current) values of the fundamental and harmonics.

An advantage of the wave analyser method of distortion measurement is that noise within the full passband does not contribute to the readout. This is because the equivalent noise bandwidth of the filter is very narrow, down to about 7Hz in some top quality (and hence costly!) instruments. Thus, relative to the maximum effective bandwidth of the amplifier, the equivalent noise bandwidth of the analyser is only a small fraction, and the noise power in the readout circuit is reduced in the same ratio.

If the equivalent noise bandwidth of the wave analyser is, say, 10Hz and the maximum effective bandwidth of the amplifier 30kHz, then the noise power in the readout circuit would be reduced by about 35dB.

Accurate measurement of THD with a distortion factor meter becomes impossible when the distortion falls about 3dB below the full-spectrum noise, but the measurement is possible with a wave analyser owing to the much smaller equivalent noise bandwidth.

Harmonic distortion measurements do not usually include hum or its harmonics, and most instruments embody a switchable high-pass filter which swiftly rolls off the response at the bass end. Hum will be included when distortion measurements are made at low frequencies, for then it is necessary to switch the filter out, of course. A note should then be made that the low-frequency distortion includes hum. This is applicable mainly to measurements based on a distortion factor meter.

**Intermodulation Distortion**

IMD refers to the interaction of two or more signals which, owing to amplifier nonlinearity, produces new sum and difference signals. When there are two signals of $f1$ and $f2$, the percentage IMD is referred to the highest one $f2$, and is expressed as

$$\frac{\sqrt{[(E_{f2-f1} + E_{f2+f1})^2 + (E_{f2=2f1} + E_{f2+2f1})^2 + (E_{f2-3f1} + E_{f2+3f1})^2 + \dots]}}{E_{f2}} \times 100$$

where $E_{f2-f1}$, for example, is the voltage of the intermodulation component at the frequency $f2 - f1$, measured at the output. When convenient, the currents of the components can be used in the calculation instead of the voltages.

Fig. 1.6 Rogers Distortion Factor Meter. This reads down to 0·1% f.s.d. in seven switched ranges.

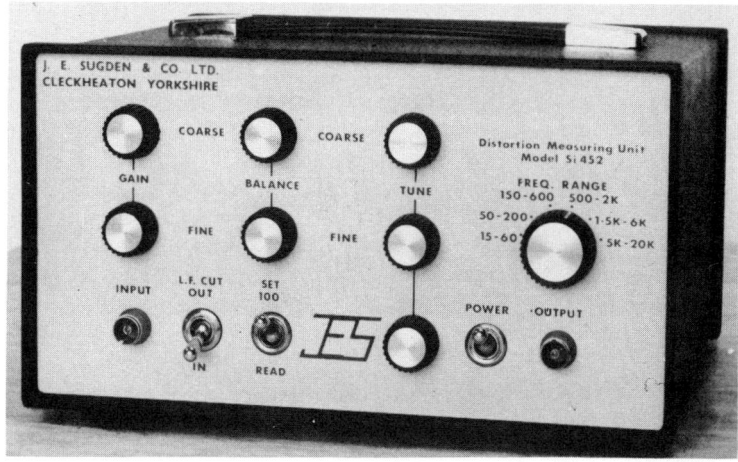

Fig. 1.7 J. E. Sugden Distortion Measuring Unit Si 452. This requires a millivoltmeter for readout.

The IMD test set usually incorporates a readout device and two oscillators for producing the signals at $f1$ and $f2$. Signal $f1$ is commonly 100Hz and signal $f2$ about 5kHz. The most used test adopts signals in the amplitude ratio of $f1:f2 = 4:1$.

As a result of amplifier non-linearity, the signal at $f2$ is modulated by the signal at $f1$ and its harmonics, thereby yielding intermodulation products of $f2 - f1, f2 + f1, f2 - 2f1, f2 + 2f1, f2 - 3f1, f2 + 3f1$. etc.

The intermodulation products can be measured individually and the percentage intermodulation calculated from the expression above. An alternative approach (Chapter 2, page 52) is based on the filtering of the intermodulation products and the measurement of their r.m.s. sum.

The output power at the measured level of IMD for the $E_{f1}$ and $E_{f2}$ components is

$$\text{Power} \quad W = \frac{(E_{f1} + E_{f2})^2}{R},$$

where $E_{f1}$ and $E_{f2}$ are the voltages of the components at frequencies $f1$ and $f2$ respectively, and $R$ the load resistor in ohms. When the current ($I$) of the components is used, then the power output is equal to

$$W = (I_{f1} + I_{f2})^2 \times R.$$

One advantage of IMD measurement over THD measurement is that oscillators of very low distortion are not necessary. However, depending on the IMD test method adopted, and on the frequencies and relative amplitudes of the signals employed, wide variations in readout or calculated IMD are possible.

Nevertheless, IMD measurements are held by some authorities to provide a greater correlation to the 'sounding' of an amplifier, particularly from the point of view of Class B crossover distortion, than THD measurements. It is certainly true that there is no simple relationship between the two.

### Frequency Response (Filters, etc)

While the power response is plotted with the amplifier towards full power, the frequency response is plotted at a nominal low power, a common value being about 1W for hi-fi amplifiers.

Instruments required are a sinewave source, adjustable and accurately calibrated over the frequency spectrum from 20Hz to 20kHz at least, amplifier dummy load resistor(s) and an audio voltmeter (e.g. millivoltmeter) with an open dB scale at the datum point.

For very accurate plots a switched attenuator with absolute calibration settings is required to avoid the interpolation of meter readings between dB calibration marks. Various changeover switches may also be required.

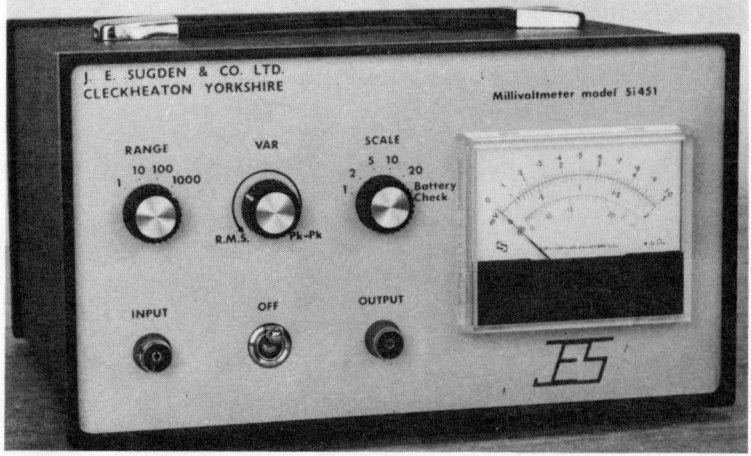

Fig. 1.8  J. E. Sugden Audio Millivoltmeter Si451. A variable control allows the readout to be adjusted from r.m.s. to peak-to-peak.

Fig. 1.9 Radford Low Distortion Audio Oscillator. This is one of the lowest distortion oscillators available (down to 0·002% at 1kHz and 0·005% from 200Hz to 20kHz).

**Input Sensitivity Voltage**

For this measurement it is necessary to know the r.m.s. input voltage at 1kHz when the amplifier is delivering its rated power with the volume control at maximum, tone controls neutral and filters out. Thus, in addition to the meter and load for measuring output power we require some means of measuring the input sine-wave signal.

Some audio generators are equipped with a meter indicating the output signal voltage (Fig. 1.9), while others have calibrated attenuators. However, if the generator fails to indicate the r.m.s. output voltage exactly (Fig. 1.10), the signal fed to the amplifier must be measured with an external audio millivoltmeter (Figs. 1.8 and 1.11).

Note that some amplifier inputs, such as those for magnetic pickup, tape head and microphone, might not require much more than 1mV for the rated output, so the millivoltmeter must read down to such low values.

**Preamplifier and Recording Signal Levels**

Preamplifier outputs and those outputs (RCA 'phono' and DIN sockets) delivering signal for tape recording also need a millivoltmeter for measurement and a sinewave input source which can be measured, since the outputs are generally referred to the input sensitivity voltage at 1kHz.

### Input Amplifier Overload

It is sometimes necessary to discover the 1kHz signal level that a preamplifier, notably the magnetic pickup equalised preamplifier, can accommodate before the onset of severe overload distortion. This test requires an oscilloscope for detecting waveform clipping, a sinewave input source with output attenuators and level control and an audio millivoltmeter for accurate measurement of the signal fed to the input.

The overload is generally referred (as a dB ratio) to the input sensitivity voltage at 1kHz, but care must be taken to ensure that the signal in the control section and power amplifier is insufficient to result in overloading here. For this reason it is desirable to monitor the output signal at the tape recording socket. If the overload is to be referred to a given value of THD, then a distortion test set will also be required.

### Hum and Noise (i.e. Signal/Noise Ratio)

This is not the same as the noise factor of an amplifier, for which test is required a noise generator (employing a saturated diode, for example). It relates to the summed hum and noise at the output when the amplifier is under full bandwidth-gain and when the selected input is shorted, loaded correctly or open-circuit.

It is common to measure the hum and noise voltage across the load with a millivoltmeter (or microvoltmeter when the hum and noise signals are very small) and express it as a dB ratio relative to the audio voltage across the same load when the amplifier is driven to its rated power at 1kHz.

It is necessary to employ an attenuator calibrated in dB steps for the most accurate results. Weighting may be applied (Fig. 1.2).

### Channel Separation

This refers to how well the two stereo channels of an amplifier are electrically isolated. Perfect isolation is not possible, so we get crosstalk between the channels, and the amount of crosstalk is expressed as a dB ratio.

One channel is driven to full power (requiring the instruments to measure this), while the other channel is correctly terminated but

## Ears and Test Instruments

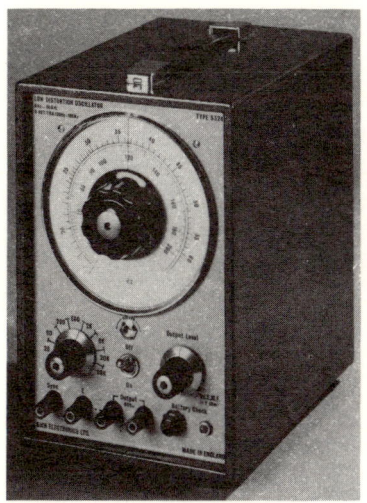

Fig. 1.10 (left) Rogers Low Distortion Oscillator, the THD being 0·05% from 100Hz to 10kHz, with second harmonic predominating. Fig. 1.11 (right) Rogers Audio Millivoltmeter, covering 1mV to 300V f.s.d. in twelve ranges.

Fig. 1.12 Corner of test laboratory, showing the Radford and some of the Sugden equipment in operation. Other instruments include Avo Electronic Test Meter, Grundig Millivoltmeter, Taylor F.M. Generator, Cossor Wobbulator, Eagle Multirange Meter and oscilloscope.

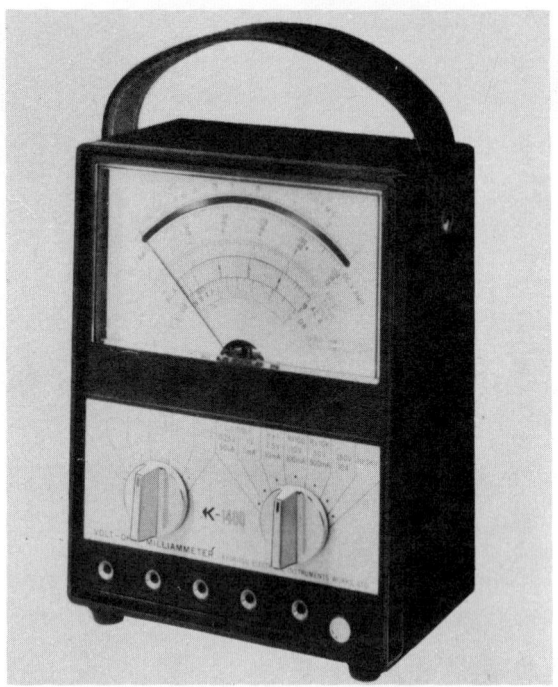

Fig. 1.13  Eagle K1400 Multirange Testmeter.

not driven. A millivoltmeter is then required to measure the signal level in this non-speaking channel.

A common source of crosstalk is the balance control, since even when this is set to perfect balance a small impedance exists in shunt with it owing to the action of the wiper on the track, and this is sometimes sufficient to introduce signal into the non-speaking channel, the level of which, of course, is very small.

**Crosstalk (Between Inputs)**

When two or more active sources are applied simultaneously to an amplifier, crosstalk from that or those not selected may appear on the selected signal. Measurement is the same as for channel separation.

Few contemporary amplifiers suffer from serious crosstalk of this kind because the inputs not selected are generally shorted by the selector switch.

## Transient Performance

Pulsed-sinewave or squarewave signal or a single stepped-pulse signal is required for this test, but the use of a squarewave is favoured by many technicians. This is relatively easily obtainable, and many sinewave generators can be switched to yield this sort of signal at any frequency within (and not uncommonly outside) the audio spectrum.

An oscilloscope is used to display the output signal (Fig. 1.3), but to observe the transient performance under conditions other than purely resistive loads, a range of capacitors and small inductors is necessary to stimulate loudspeaker loading. Capacitor, inductor and resistor boxes are thus useful ancillary items.

The performance is tested up to full power, so some method of measuring the output power must also be available.

When a squarewave is used the rise time must be very small, but modern instruments are capable of producing signals with a rise time as small as $0.1\mu S$.

## INSTRUMENTS FOR TUNER TESTS

### Frequency Range

The instrument required for this is an accurately calibrated r.f. signal generator covering Band II, as well as the long, medium and short wavebands when the tuner or tuner section of a tuner–amplifier includes them. A generator with crystal check points is useful, but instrument calibration can usually be checked against a broadcast station of known frequency.

The output will need to be monitored, but if the generator is modulated the resulting tone from the loudspeaker is sufficient.

### Sensitivity and Signal/Noise Ratio

The most important requirement for this test is a signal generator with a very accurately calibrated attenuator, with very good matching from the attenuator to the tuner or tuner section under test and with very little radiated signal field.

The latter is necessary especially when tests are performed on the latest f.m. tuners, etc., for the sensitivity of these is often such that almost full response can be obtained merely from the radiated field or via leakage signal past the attenuator! Mismatching also puts the attenuator in error. The inexpensive type of general servicing instrument is thus totally inadequate for meaningful tests of this kind.

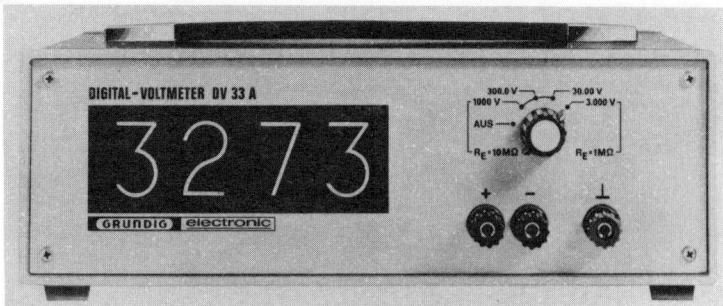

Fig. 1.14 Grundig Digital Voltmeter DV33A. This kind of instrument is useful for audio testing and for 'counting'.

However, if one has a good, well screened generator with a poor attenuator, it is possible to use the signal from the generator in conjunction with a detached, accurate attenuator, the attenuator then possibly costing more than the generator.

A readout instrument for hum and noise measurements, relative to the full output of the tuner or tuner–amplifier (with the volume control in the latter case suitably positioned) is also required. The generator must also be capable of modulation, and it is useful if the modulation depth can be adjusted, particularly the deviation on f.m.

**Frequency Response**

The most accurate way of checking this, especially with f.m. tuners and tuner–amplifiers carrying de-emphasis, is by using a generator whose modulation signal can be applied from a separate audio generator, via suitable pre-emphasis. The frequency response, taking in the effect of the de-emphasis, can then be checked with properly modulated signal from aerial input to audio output.

Alternatively, the audio signal from a generator can be applied across the detector load and monitored at the audio output, but extreme care has to be taken with this method since incorrect loading of the generator to the input circuit can result in serious error of readout, especially when a stereo decoder follows the f.m. detector.

The measurement of stereo frequency response really calls for a stereo (multiplex) generator whose left and right channels can be modulated with external signal from a separator audio generator (or generators). However, some stereo generators carry their own left and right modulating signal sources.

## Capture Ratio (F.M.)

This is a very tricky measurement, which to be meaningful requires two signal generators and star network for combining the two signals to the tuner aerial input. More is said about this in Chapter 3.

## Selectivity

There are two methods of measurement, one using a single generator and the other (better) using two generators, one of which needs to be modulated and accurately tunable over small frequency intervals. The selectivity is often given as a dB ratio relative to a specified signal/noise figure, so the measurement also requires a millivoltmeter or similar instrument for readout.

## Image, I.F. and Spurious Response Rejection

Measurements of this kind also require one or two signal generators, one to produce the wanted signal and the other the unwanted signal, each with switchable modulation, when the two generator method is adopted. The generators, as with the previous tests, need to be accurately calibrated and easily tuned over small frequencies. They should also incorporate accurate attenuators, or external attenuators can be used.

Again, a method (suitable for the tuner or tuner–amplifier) is required to measure and compare the levels of the output signals.

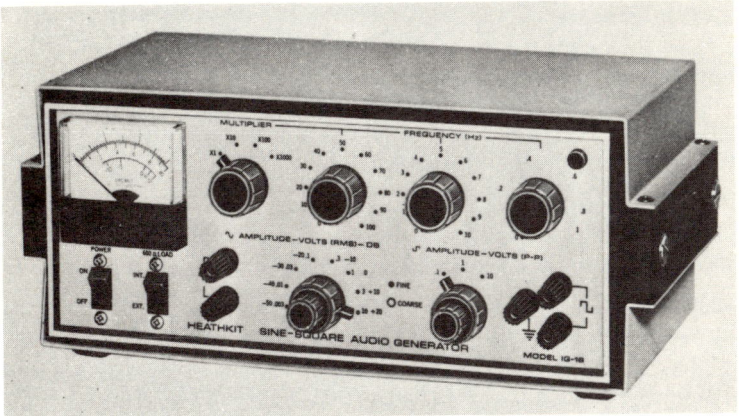

Fig. 1.15  Heathkit Sine-Square Audio Generator Model IG-18.

## Harmonic Distortion

This requires a signal generator tunable over the required broadcast band or bands which is modulated by a generator of very low distortion. The audio output is monitored at any suitable level, including that corresponding to the rated output at full modulation, and a distortion test set is used to measure the distortion in the signal.

A signal generator which will accept external modulation signal at a given or adjustable depth or deviation (in the case of f.m.) is most suitable, for then the modulation signal at any frequency (provided the generator modulator will handle it without distortion) can be applied from a low distortion oscillator. Remember, though, that noise and residual hum (unless the latter is filtered) on the r.f. carrier wave will be measured by a simple distortion factor meter along with the THD.

Distortion checks on stereo tuners or tuner section require a stereo (multiplex) generator to make the stereo decoder active and then each channel (in turn) of this is modulated with a low distortion sinewave signal of suitable frequency.

## Stereo Separation

This test also requires a stereo-encoded signal with modulation on one channel (switchable over the two channels). At a given level of modulation and relative to an established output datum, the crosstalk in the non-speaking channel then being referred to this and expressed as a dB ratio. Tests are usually made at 1kHz and 10kHz, the latter because as the modulation frequency rises the crosstalk tends to increase.

Alternatively, the stereo test transmissions radiated from stereo Radio 3 stations of the BBC after late-night close-down can be used with reasonable accuracy for left-to-right and right-to-left crosstalk measurements and for other tests when such a station yields adequate signal field at the test site (see BBC Engineering Information Sheet No. 1605). See the table, page 99.

## Stereo Switching and F.M. Muting Levels

F.M. tuners and tuner sections carrying stereo decoders often have automatic switching to stereo which is activated by the pilot tone of the resulting 38kHz reference signal. The v.h.f. signal level at which this switching occurs can be measured with a stereo generator having an accurate attenuator. The switching is generally indicated by the glowing of a stereo indicator lamp.

*Ears and Test Instruments* 29

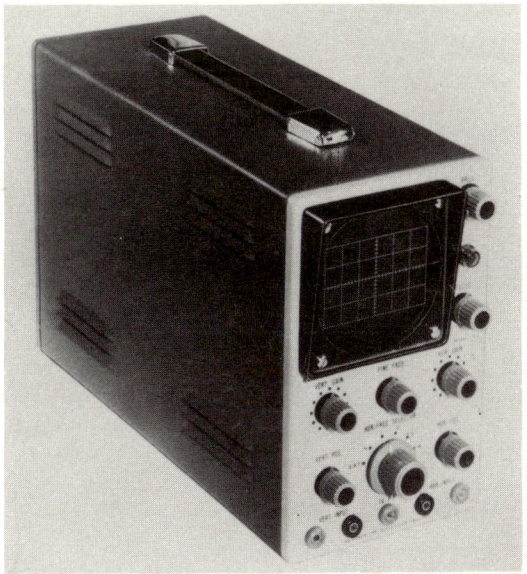

Fig. 1.16 Inexpensive oscilloscope (OS-2) by Heathkit, suitable for audio tests.

In a similar manner the switching point of the f.m. interstation muting can be measured, but this requires an ordinary v.h.f. generator, with or without modulation, carrying an accurate attenuator.

## INSTRUMENTS FOR OTHER TESTS

The instruments mentioned so far will permit exhaustive testing of amplifiers and tuners, and will also go a very long way in the testing of other audio equipment, including gramophone pickups, turntable units (for rumble, etc.), loudspeaker, headphones, tape recorders and so forth.

As is the nature of our craft, though, each item of equipment calls for its own range of specialised test equipment. However, unless it is one's job to test in great detail each parameter of the specification (and more) or to design and develop a piece of audio equipment, then we can generally get away with a modest display of test kit provided we know how to improvise and can employ each piece of test equipment to its maximum.

Nevertheless, we shall need plenty of ancillary items, such as attenuator pads, decade attenuators, high wattage load resistors, capacitor, inductor and resistor boxes, plugs of all kinds, adaptors of various kinds (DIN to 'phono', and vice versa, for example), and so on.

If tape recorders come within our province, then we shall need a host of test tapes carrying a variety of test signals. We might also find a wow and flutter meter and companion wave analyser for examining the wow and flutter components useful.

Although some items of test equipment incorporate a filter or filters (such as the high-pass filter in a distortion factor meter), there are often times when we need to include a high-pass, low-pass or bandpass filter in a test setup to remove hum, noise or to allow the passage of a specific band of frequencies.

It is possible to utilise the filter or filters of the test equipment in many cases, but when this is not a feasible proposition external filters have to be employed. Thus, the ancillary items of practising audio technicians and technician-engineers include switchable and tunable filters with suitable matching devices.

**Test Discs and Tapes**

For gramophone pickup and turntable unit testing we require a range of test records to produce the various test signals and com-

Fig. 1.17 J. E. Sugden Low Distortion Oscillator Si453. THD is about 0·01% at 1kHz. The instrument is battery operated and is complementary to the Si451 and Si452.

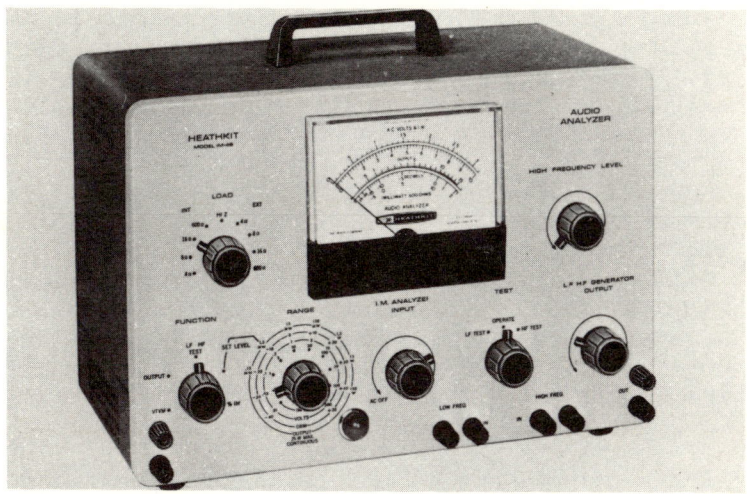

Fig. 1.18 Heathkit Audio Analyser Model IM-48 designed for intermodulation distortion measurement.

parative signals. Some useful ones are *EMI* TCS101, TCS102, TCS104 and TCS105, and for tracking tests TS201; *Hi-Fi Sound* HFS69; *CBS* various specialised frequency test discs, including the well-known STR100.

Test tapes include *EMI* TBT.1A and TBT,2A and the *Tutchings Electronics Ltd.* range. There are hosts of others, including test discs, with additions being made for specific purposes or for updating quite frequently.

**Bread and Butter Instruments**

As already intimated, a workshop equipped with the instruments so far mentioned would be well able to handle the vast majority of servicing procedures, in addition to detailed testing.

Such a workshop would almost certainly also be well equipped with the basic 'bread and butter' instruments, including multirange testmeters, electronic testmeters (i.e. valve or transistor voltmeters), resistance and impedance bridges, stabilised power supplies, a wobbulator for visual alignment in partnership with the oscilloscope, etc.

There is no need in a book of this kind to delve deeply into the nature of instruments like these, since most technicians use them every day and are thus very well conversant with them.

However, a word or two about the oscilloscope would not be amiss. For audio work this should have a Y bandwidth of several MHz (7 to 10MHz if it is to be used also for colour television servicing) and the Y sensitivity should be at least 100mV/cm. Triggered and repetitive sweeps should be available and the sync should be stable.

A design suitable for accommodating a camera is useful when oscillograms (perhaps for publication) might be needed. There should also be X expansion and facilities for calibration.

More audio technicians and technician-engineers are finding the latest digital instruments useful. Already there are various species of multirange testers of this type. They provide very accurate readouts, and some digital voltmeters can second as counters, useful for accurate frequency determination.

CHAPTER TWO
# AMPLIFIER TESTS

It is not intended in this and the following chapters to explain the elementary servicing procedures, for technicians and technician-engineers dealing with audio equipment will be adequately conversant with these from their work on radio and television servicing. Nevertheless, we shall be investigating various fault symptoms and conditions and the tests required to reveal their whereabouts.

**Basic Functions**

An audio amplifier provides three basic functions. One, it accepts and equalises the programme signals, while at the same time amplifying them. Two, it provides various controls over the signals and in some cases filters them. And three, it translates the resulting small power signals to large power ones for the purpose of driving a loudspeaker system.

There are other operations, of course, which are related to the above mentioned functions, including the selection of the programme signals, the provision of low power signals for tape recording, the provision of medium power signals for working a headphone set, the differential control of the gain of the two channels in a stereo amplifier for the purpose of balancing, the switching of two loudspeaker systems and the headphone circuit, the switching of the various filters and sometimes frequency and slope adjustment, the switching of programme signals prior to the power section for mono and stereo mode, the switching for tape monitoring, etc.

There is no typical amplifier as such, although a basic pattern is followed by the manufacturers of most of those designed for hi-fi applications. Most of them nowadays, however, are designed for stereo operation, meaning that there are two isolated channels with switching as already mentioned. The two channels are integrated into a common housing and usually supplies the power unit for both channels.

### Three Sections

It is convenient to regard an amplifier under test as being composed of three sections which provide preamplification and equalising (where necessary), control and secondary amplification and power amplification.

In addition to these sections, of course, is the power supply unit, and in some designs this in itself can be quite complex, since it might incorporate supply regulation, current overload protection and stabilised, low impedance feeds to the various amplifier departments. The power amplifier might also have a drive overload protective circuit tied up with it.

The majority of contemporary amplifiers are of the so-called 'integrated' variety, which merely implies that the power amplifier is built into the same housing as the equalised preamplifiers and control section. Early amplifiers of the valve era were commonly in two sections, one carrying the equalised preamplification control and secondary amplification, and the other the power amplification and power supply unit. Thus the two sections had to be coupled together both for power supply and signal continuity.

### Non-integrated Amplifiers

We still come across solidstate amplifiers like this when the amplifier power is particularly high and when the design is specialised. There are certain advantages to be gained from this non-integrated design, one being that the two sections can be operated completely independently when each has its own power supply, which is generally the case with solidstate equipment.

Some power amplifiers are designed for hidden operation (at the base of a large loudspeaker enclosure, for example), the control section signals then being fed to it through screened cables at relatively low impedance (from an emitter-follower). Thus the control section would be that accessible for the user. Others, though, have controls and switches and need to be placed accessibly with the control section.

Fault finding and testing are certainly easier to handle when the design is non-integrated and each section fully independent, for each can then be worked on as a separate item of equipment. However, there is a trend (particularly so far as Oriental imports are concerned) for the two sections to be arranged for independent operation even when the design is integrated.

This is based on the output of the control section being terminated to sockets, and likewise the input of the power amplifier section. Under normal operating conditions wire links or locked switches are used for signal continuity, but when independent operation is required the links are removed or the switches unlocked and manipulated.

**Power Capacity**

A primary test of any amplifier is power output at 1kHz and at frequencies at the bottom and top ends of the spectrum. 40Hz and 10kHz are the extreme frequencies sometimes chosen, the power bandwidth (see later) then giving an idea of how the power falls at probably even more extreme frequencies.

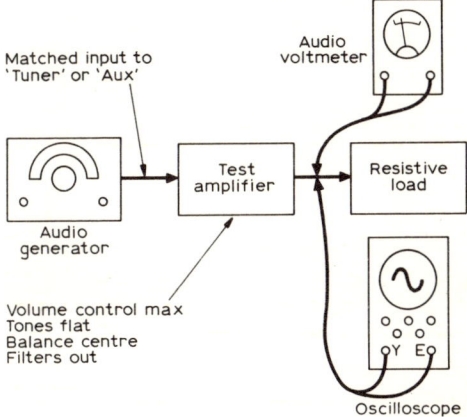

Fig. 2.1  Instrument setup for power measurement.

The most meaningful power is that equal to $I^2R$ and $V^2/R$, where $R$ is the resistive load in ohms. $I$ and $V$ are r.m.s. current and voltage. The test setup is given in Fig. 2.1, and when the amplifier is integrated the input is applied to a selected programme signal socket, usually tuner or auxiliary. It can be likewise applied when the amplifier is non-integrated, but in this case the signal might be more conveniently applied to the power section input.

When the control section is in circuit, however, the volume control should be set to maximum, the balance to 'neutral', the tone controls to 'flat' and all filters switched out.

Input signal amplitude is gradually increased until the peaks of the waveform start to clip as shown in Fig. 2.2, and then the input is turned back slightly until the clipping just disappears. At this setting the r.m.s. voltage is read off the audio voltmeter and the power calculation made (a slide rule is useful for this). This is the *power capacity* of the amplifier at the particular frequency tested.

**Rated Output**

The *rated* and *maximum* output power might be slightly below the power capacity since with the latter the threshold is waveform clipping, while with either of the former the power is referred to the rated THD of the amplifier. For this test, therefore, a distortion meter is also required (see under THD).

The maximum output power, however, might differ from the rated output power because the THD at the rated power might be below the rated THD, in which case a greater output (the maximum) would be obtained by increasing the input signal amplitude until the rated THD value is achieved. Note that the THD could be well above the rated THD at waveform clipping threshold.

An amplifier should deliver the rated or maximum output power continuously without distress for a period of at least 30 seconds at 1kHz, and most with the 'hi-fi' label are capable of this.

**L. F. Limitations**

Since the power *capacity* at the l.f. end of the spectrum is often limited by the onset of severe waveform distortion (sometimes a shortcoming of the coupling capacitance or negative feedback), the power is best related to a given value of THD, but this in terms of a *rated* value is not always given in the specification at l.f.

There is not usually any trouble in obtaining a rated or maximum power measurement at h.f.; but in poor designs crossover distortion might be visible on the waveform at 20kHz. The THD of some of the best designs minimises at around 8–10kHz.

**Stereo Amplifiers**

With stereo amplifiers each channel should be measured for power delivery separately and the difference should not exceed about 1dB. It is instructive, however, to drive both channels together and then to compare the per-channel power with that obtained when only one channel is driven. The quickest way of doing this is to apply the generator signal to, say, a left input and then to switch the control

## Amplifier Tests

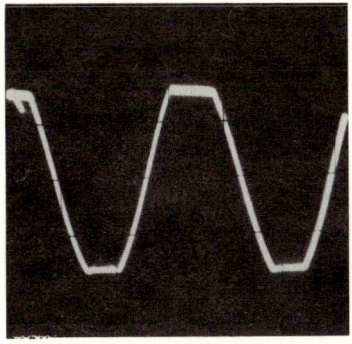 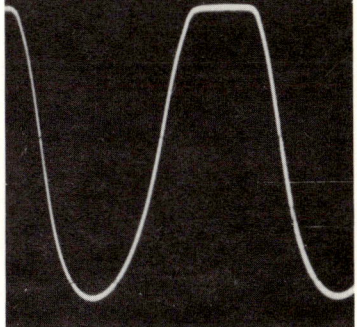

Fig. 2.2 (left) Sinewave peak clipping.   Fig. 2.3 (right) Asymmetrical clipping.

section to mono. Both power amplifiers will then be driven by the common generator signal.

This test reveals unduly high impedance in the power supply feeds and poor supply regulation in Class B designs. When both channels are wound up to full power, of course, the power supply is called upon to deliver almost twice the single channel power, and as a consequence the supply voltage falls when the regulation is poor.

Those amplifiers with very good regulators barely show any difference in per-channel power with one or both channels driven, but amplifiers of lesser exactitude in this area can drop by almost 2dB when both channels are driven together. A common value is 1dB at 1 kHz and 8Ω loads. Class A amplifiers, since their input power is theoretically constant over the entire signal cycle, are not so affected by two-channel drive.

**Instability Tendencies**

Because there are some resistive or impedive supply circuits common to both stereo channels, l.f. instability tendencies are emphasised when both channels are driven to maximum together by the development of a spurious signal, commonly at very low-frequency, causing the displayed waveform to pulsate vertically at low repetition rate.

If this trouble develops over the life of an amplifier attention should be directed to the power supply filtering components, such as large electrolytics, zener diodes, regulator transistors and associated components.

When testing at h.f. into low value loads and towards full power, the protective device may persist in functioning and cutting out the power transistors or the d.c. supply fuses may blow. This is not abnormal, since the current in the power transistors at h.f. may rise above that at 1kHz and below owing to hole storage effects and delayed switching phenomena in Class B designs. The action of the negative feedback can also have a bearing on this.

**Dummy Loads**

Power amplifier channels should be loaded by pure resistances—not with the loudspeaker systems, even if you and the loudspeakers could stand the sustained audio power! This is because loudspeaker systems look reactive to the amplifier and they could reflect their electromagnetic characteristics into the tests. The same applies to distortion and transient tests.

Since we are dealing in continuous power, the loads will gain temperature quickly, and so must be man enough to handle the full steady-state power without changing resistance unduly or getting red hot and fusing. Large wirewound resistors are thus essential, and it is often necessary to heat-sink these to dissipate the heat so that the resistance does not rise too much during the tests (assuming resistive elements of positive temperature coefficient).

Since we need to test the power at various load values it is a good idea to construct a resistor box with switching permitting high-wattage loads from 4 to 16Ω easily to be applied to the amplifier output terminals. A load-off position on the switching is also useful when measuring the damping factor or source resistance (see later).

As mentioned in Chapter 1, a rheostat of suitable power rating makes a good adjustable-value load provided the resistance can be measured accurately or the rheostat is calibrated in ohms. Heat-sink loading of these also helps to up-rate the device and keep the temperature reasonable during sustained operation.

The power based on r.m.s. voltage or current is the continuous or steady-state power. The *peak* power is just double this since it is based on peak voltage or current which is $\sqrt{2}$ the r.m.s. value—hence the doubled power rating, resulting from the 'square' function in the power expression.

Other methods of power measurement are geared more towards a simulated music waveform (Chapter 1), as distinct from a steady-state sinewave signal. There are various schemes, including that by IHF (see IHF-A-201:1966 under paragraph 3.0). However, all the tests that we need to perform are best related to continuous power (BS 3860:1965 under paragraph 10).

## Amplifier Tests

Fig. 2.4 Inside view of Class B amplifier showing the mid-point, voltage and quiescent presets for each channel.

## Low Power

An amplifier not delivering its rated power would almost certainly be in trouble in the power amplifier section or power supply. Of course, low sensitivity of the preamplifiers would reduce the power amplifier drive and give the same effect when the input is equal to the sensitivity voltage. However, when testing essentially for power one rarely takes particular heed of the overall sensitivity since the generator signal is progressively increased (overcoming any sensitivity shortcoming) until the output waveform is seen to clip (Fig. 2.2).

Thus it is best to investigate low power with reference to waveform clipping level. A common cause of failure to realise the rated power is clipping asymmetry of the power amplifier. What happens is that the peak of one cycle clips before the other, as shown in Fig. 2.3. In practice, this does not reduce the power yield very much when the asymmetry is only mild; but in extreme cases it can reduce the rated power by about 2dB.

## Drive Preset

Most amplifiers are equipped with a preset control to balance the drive half cycles. VR1 in Fig. 2.5 is such a control, and this is set at *full output* for equal peak clipping. Its purpose is to set the mid-point voltage of the output pair TR6/TR7, and this it does by the change in d.c. conditions reflected from the change in TR2 base bias, via the d.c. couplings.

This is a fairly common arrangement for obtaining symmetrical clipping, and the presets involved (one for each channel) can be seen in the photograph in Fig. 2.4. This shows *two* presets each side of the mains transformer. Two are for one channel and two for the other. One of each pair adjusts the mid-point voltage for symmetrical clipping as we have seen, while the other is concerned with quiescent current adjustment, considered later.

Similar presets can be seen in Figs. 2.6 and 2.7. Sometimes there is only one preset per channel for quiescent current adjustment, the mid-point voltage then being established by fixed resistors in the

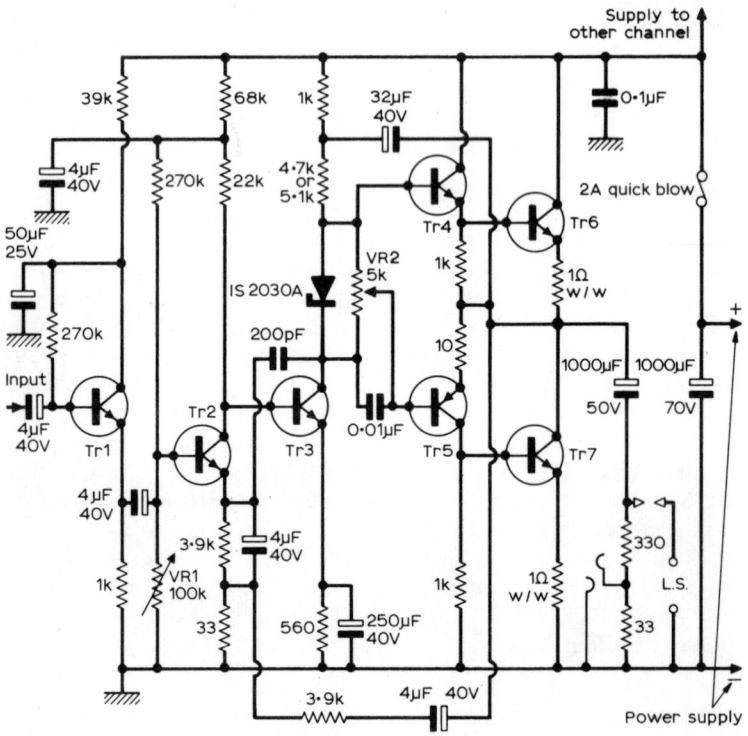

Fig. 2.5 Circuit of quasi-complementary power amplifier, where TR4/TR5 form the complementary pair driving TR6/TR7 n-p-n devices. TR1 is the secondary preamplifier and TR2/TR3 a d.c. coupled pair of n-p-n devices working as pre-drivers. VR1 sets the mid-point voltage for symmetrical clipping, while VR2 adjusts the quiescent current of the output pair.

## Amplifier Tests

Fig. 2.6 The adjusting presets can be seen in the bottom left-hand corner of this photograph.

Fig. 2.7 The adjusting presets (pair for each channel) are clearly visible on this module.

biasing circuit. If the clipping is asymmetrical from such an amplifier there would be a possibility of the biasing resistors having altered in value.

### Power Capacity

The power capacity of a power amplifier is closely geared to the supply voltage for a given value load in accordance with the following expression for Class B amplifiers.

$$\frac{(V - v)^2}{8R}$$

where $V$ is the supply voltage, $v$ the voltage lost across the transistor and series resistance and $R$ the value of the load in ohms.

Clearly, the smaller the value the load or the greater the supply voltage, the greater the power capacity. Maximum power, of course, is limited by the transistors, but the expression shows that if the supply voltage falls even mildly the power will fall significantly.

As much as 2·5dB can be lost by running an amplifier which has no voltage regulator on the incorrect mains voltage tapping. The converse will apply, of course, if the tapping is set too low relative to the mains voltage. Hence it is essential to ensure that the mains tapping matches as closely as possible the applied mains voltage. A constant voltage source is desirable for serious tests, but the waveform from this should not deviate unduly from pure sinewave.

### Voltage Regulators

Amplifiers with voltage regulators are not prone to this effect so much, but if the maximum power falls below the rated power a voltage check should be made to ensure that the regulator is delivering the specified voltage. The circuit of simple series regulator is given in Fig. 2.8.

TR1 is the transistor through which the supply current flows (from collector to emitter), and the output is dependent on its conductivity. This is regulated by the base bias which is derived from TR2, via the d.c. coupled TR3, the base of which monitors the supply voltage, via VR1.

Thus, if the supply voltage falls due to an increasing load, TR1 is turned on more for increasing conductivity so that the load demand is satisfied. The converse applies, of course, should the supply voltage tend to rise due to a decreasing load.

VR1 regulates the nominal conductivity of TR1, via the loop of the d.c. coupled pair, so by adjusting this preset the main rail voltage is altered about the value required by the power amplifier for maximum audio output.

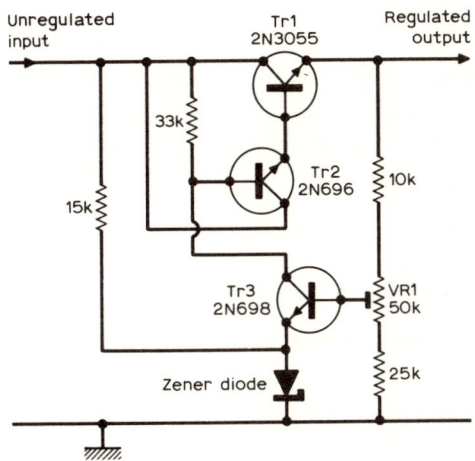

Fig. 2.8  Circuit of simple series regulator.

## Zener Reference

The zener diode in TR3 emitter circuit stabilises the control potential and provides a 'reference', while in a circuit like this the unregulated input might be about 70V and the regulated output about 60V or a little less.

Such a regulator has a very low effective source resistance and thus assists with the supply filtering at low frequencies, down to d.c. It also makes the amplifier less susceptible to mains-borne transients as are often produced by the switching of electrical equipment in the domestic scene. RC filter circuits are sometimes incorporated to protect the regulator transistors themselves from mains-borne transients.

## Overvoltage

Some imported equipment is reaching the British market with a mains input of 220V nominal only. On 240V supplies the rail voltage might be in excess of that required for the amplifier to deliver its rated output. Thus when such amplifiers are tested the audio delivery might be significantly above normal.

Well-designed species appear not to suffer from this overvoltage when working normally, but if they are driven towards full power without an output load there is a possibility that the driver or output transistors (or both) might fail, due to excessive collector-emitter voltage on the swings.

## TOTAL HARMONIC DISTORTION

The setup required for measuring the THD is given in Fig. 2.9. Two oscilloscopes are shown here, one to display the waveform across the load, prior to the THD test set, and the other to display the waveform of the THD and noise residual.

In practice, one oscilloscope is sufficient since it can be connected from one source to the other, as required, or both signals can be monitored simultaneously when a dual-trace oscilloscope is used. This can be a dual-beam (or split-beam) model or a single-beam instrument used with a 'trace doubler'.

The heart of the distortion factor meter is a deep, narrow notch filter which is tuned to 'notch out' the fundamental frequency, leaving a signal composed of all the harmonics and noise over the amplifier's bandwidth. This signal is measured by the audio millivoltmeter and its r.m.s. value is given as a dB ratio or percentage relative to the r.m.s. value of the voltage of the signal direct from the amplifier across the load, and for this reason the readout meter is calibrated either in dB or percentage.

It is important to remember that when converting dB ratios to percentages it is the voltage—not the power—ratio which is used. For example, 40dB is a 100:1 voltage ratio which, of course, is 1%. Similarly, 60dB is a 1,000:1 voltage ratio, corresponding to 0·1%. and so on.

### Attenuator

Distortion measuring instruments usually incorporate an accurate attenuator which, in conjunction with a 'fine' control, is adjusted to establish a datum corresponding to the r.m.s. voltage of the test signal across the load.

The notch is then switched in and tuned and the attenuation is switched down in suitable steps effectively to increase the readout sensitivity so that the meter will indicate the r.m.s. signal due to the distortion components.

Stages of fine tuning are also incorporated so that the tuning of

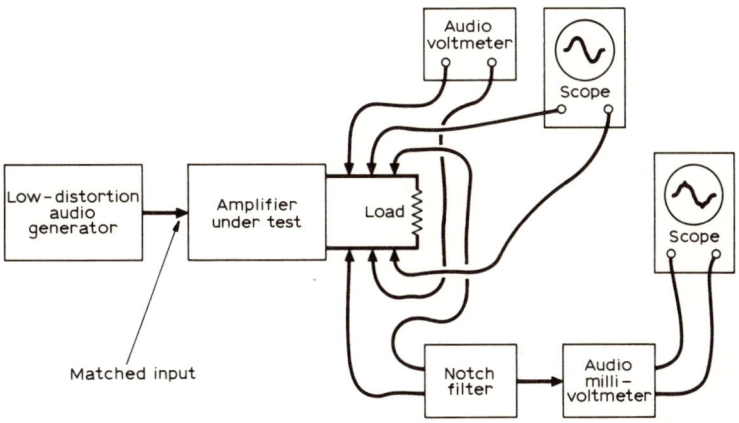

Fig. 2.9  Instrument setup for THD measurement.

the notch can be optimised each time a step reduction of attenuation is introduced. The switched attenuator may be calibrated in terms of percentage scales on the readout instrument or in decibels, the meter scale or scales then being likewise calibrated.

The plan is to adjust the notch tuning for the least reading, which results when the fundamental frequency of the output signal is attenuated to the maximum possible by the nature of the notch filter design (this usually being in the order of 80 to 100dB).

For correct readout, however, it is also necessary to 'balance' the distortion waveform, and for this various stages—going from 'coarse' to 'fine'—of balancing adjustments are provided. Waveform asymmetry is introduced by the action of the notch filter, and unless this is carefully balanced out each time the notch tuning is adjusted, the THD readout will be considerably greater than the actual THD present.

**Residual Noise**

Now, because the readout circuit is wide open to the full treble bandwidth of the test amplifier, the residual noise of the amplifier is bound to be indicated by the readout meter, meaning that the meter will indicate in quadrature sum the THD and the noise. However, this only becomes troublesome when the level of the THD approaches that of the noise. Thus, if the noise of an amplifier is, say, 60dB below its rated output, the minimum THD that can be measured relative to the rated output is obviously 60dB, which, as we have seen, is 0·1%.

This may be fair enough, but if we decide to measure the THD with a distortion factor meter at, say, 20dB below the rated output, then at that output—assuming that the noise condition of the amplifier is unchanged—we would only be able to measure down to $-40$dB of THD before reaching noise, which is 1%.

This is certainly not so good because nowadays we often require to discover whether the distortion tends to increase or decrease as the power yield of the amplifier is diminished. It will be recalled (see, for example, my companion volume *Tuners and Amplifiers*, by the same publisher) that some Class B amplifiers tend to increase in THD as the power is reduced, which is the converse of the Class A tendency. To prove this, therefore, we have to check the THD right down to 100mW or less.

## Effect on THD Readout

If the amplifier we are testing is rated at, say, 50W, then at 100mW the power is down by about 27dB. If the noise is 60dB below the rated output, as in the previous illustration, the least distortion measurable before noise seriously starts to affect the readout would be about $-33$dB (60dB minus 27dB), which corresponds to about 2·25%. In reality, the readout would be almost totally noise signal, since the distortion would be buried well beneath it when the amplifier has a hi-fi qualification.

The margin reduces, of course, as the power is turned down. At 10mW, for example, the 50W amplifier would be running at about 37dB below rated output, meaning that we could measure 23dB THD down to noise, corresponding approximately to 7%.

Viewing the situation another way, we can say that the dB value corresponding to the noise below the rated output needs to increase by the amount of the dB power decrease in order to maintain a THD readout down to noise equivalent to that possible at the rated output.

Take a 10W amplifier whose noise is 60dB below 10W. At 10W, therefore, we can measure 0·1% THD down to noise. To measure the same percentage of distortion at 10dB below 10W, however, the noise below 10W must be 70dB, and 80dB when the output is reduced to 20dB below 10W, etc.

This function occurs quite simply because we are referring the noise to the rated output and the THD to the signal voltage across the load at the power of the measurement.

Obviously, the THD readout is bound to be affected by the noise, but the resulting readout error is small up to the point where the amplitude of the noise equals that of the THD, increasing thereafter.

## Distortion Factor

When the noise level is known, it is possible to calculate it out, leaving only the THD. A distortion factor meter readout corresponds to the ratio of the r.m.s sum of the impurity components (e.g. THD and noise) to the r.m.s. value of the total signal voltage and can be expressed as distortion factor ($DF$), such that

$$DF = \frac{\sqrt{(N^2 + D^2)}}{S}$$

where $N$ is the noise voltage, $D$ the r.m.s. sum of the harmonic voltage components and $S$ the total r.m.s. signal voltage.

The ratio $D/S$, corresponding to the THD, is thus

$$D/S = \sqrt{[DF^2 - (N/S)^2]}$$

It is often possible to measure the noise level (see under Hum and Noise) and thus calculate the exact amount of THD in the readout, but in practice accurate measurement is virtually impossible when the THD falls by about 3dB below the noise level.

## Residual Hum

Most distortion factor meters, test sets and units are equipped with an output for feeding the residual to an oscilloscope for waveform analysis, as explained and illustrated in Chapter 1. The residual of course, will also show the noise, and by using the graticule of the oscilloscope it is possible to obtain some idea of the relative amplitudes of the THD and noise, allowing a quick calculation to be made of the noise contribution in the readout.

Residual hum at 50Hz or harmonics thereof will also contribute to the readout and put the THD reading in error. To avoid this, however, distortion test sets embody a high-pass filter which deletes the hum components without affecting the near harmonics of the fundamental.

This filter, of course, can only be used when the fundamental of the test signal is above the filter turnover frequency. It cannot be used when the THD of low-frequency signals is being measured. At low frequencies, therefore, the readout may be above the real THD due to the influence of the residual hum from the amplifier under test.

## Low-noise Input

Now, in spite of the noise problems previously investigated, it is often possible to obtain THD readouts without excessive error due to noise, and without having to calculate the noise out, by applying the generator signal to an input relative to which the hum and noise below the rated output is small (i.e. a high dB value).

For example, with an integrated amplifier where the power amplifier can be used independently of the control section or with a non-integrated amplifier, the generator signal can be applied direct to the power amplifier input where the hum and noise might be 100dB below the rated output.

In this case a THD readout down to 0·1% could be obtained with the amplifier running at 40dB below the rated output. With the amplifier running at full power, the THD readout down to noise would correspond to 0·001%, which is a very small amount of distortion!

When it is not possible to connect direct to the power amplifier, then an input of the lowest noise should be selected and the generator signal applied to this. On hi-fi amplifiers the radio or auxiliary input has a hum/noise about 70dB down, which means that 0·1% THD can be measured with the output at 10dB below the rated output.

Moreover, since the volume control is often located towards the end of the control section, the noise signal at the output load can often be reduced by turning down the volume control, rather than by turning down the power at the sinewave input source.

**Noise Ratios**

We shall see later that the noise, etc., at the output load is measured in terms of a ratio to the rated power when the volume control is fully advanced. This ensures that the noise includes that generated by the preamplifiers and control unit secondary amplifiers.

Indeed, most of the output noise is amplified noise originating in the early stages. Thus we have the noise of the power amplifier to which is added the noise fed via the volume control from the pre-stages. Noise signals add in quadrature, so that the total noise $(N_t)$ is

$$N_t = \sqrt{(N_{pre}^2 + N_{power}^2)}$$

where $N_{pre}$ is the noise voltage of the early stages prior to the volume control and $N_{power}$ is the noise voltage of the power amplifier section. Clearly, then, $N_{pre}$ contribution diminishes as the volume control is turned down.

With the volume control turned right down the noise at the output load is then that of all the stages after the control, which commonly results in a ratio as high as 86dB or more. The best plan, therefore, is to measure the THD at the rated power with the volume control

## Amplifier Tests

at maximum (this technique prevents the input stages from being overloaded) by applying the sinewave input signal at a level corresponding to that of the sensitivity voltage of the selected input, and then turning down the power for THD measurements below the rated power by means of the volume control.

When the noise at the input socket selected provides a ratio of about 70dB (as it often does on tuner or auxiliary), THD at an output power of 100mW or less can generally be measured easily down to $0.1\%$ without the noise significantly affecting the results.

However, when distortion is being measured from signal passing all the way through the amplifier, the readout obviously includes the distortion of the preamplifiers, control section and secondary amplifiers in addition to that of the power amplifier section. Like noise, distortion adds in quadrature.

## Control Section Distortion

It is possible, of course, to measure the distortion produced by the control section when the design is integrated and carries a control section output socket or sockets and when the amplifier is of a non-integrated design.

The input amplifier of the distortion measuring set generally has an attenuator (stepped and 'fine') for adjusting the input signal acceptance level, so even relatively small signal voltages from the control section can be measured for distortion provided the test equipment noise is sufficiently low.

The output of the control section needs to be correctly loaded, and the remarks already given about amplifier noise apply equally when this kind of testing is adopted—possibly to a greater extent, owing to the higher gain of the preamplifier and the resulting higher noise signal.

As already mentioned, the distortion is commonly measured with the tone controls 'flat' and all filters 'out'. However, it is sometimes instructive to check the distortion at the appropriate frequencies with the tone controls set for maximum boost! The overload capability of each input stage can also be related to a given level of distortion, rather than to waveform clipping.

## Wave Analysis

A THD (or distortion factor) test is very convenient and the equipment required not all that expensive. However, for more detailed information the measurement of each harmonic separately is necessary, and for this one requires a wave analyser.

There are so many different and specialised instruments of this kind available that it is not possible to be specific in terms of application. Moreover, a good wave analyser costs in the order of £1,000, so they are rarely found in the laboratories or workshops of the service technician.

Briefly, operation consists of tuning the instrument to the frequency of the signal component to be measured, and its amplitude is then obtained directly from a fascia readout device (moving-coil meter and/or digital-meter). Before this can happen, though, various controls have to be carefully adjusted to set the frequency range, the window bandwidth, etc., while sophisticated instruments include such refinements as auto-ranging voltmeter, auto-lock, auto-frequency sweeping, etc.

**Automatic Lock**

Some analysers embody circuits which automatically tune and lock the filter. This is something like 'automatic frequency control', so that once the user has tuned approximately the instrument automatically seeks and locks on to the exact frequency of the signal component being rejected or measured.

To find the THD, of course, the r.m.s. sum of all the harmonic components needs to be calculated (the distortion factor meter does this automatically), but the distortion analyser has the advantage over the distortion factor meter of a very small equivalent noise bandwidth which, with first class models, is less than 10Hz at the −3dB points.

Thus, since the total noise bandwidth of an amplifier under test is likely to be at least 30kHz, the noise power in the readout channel would then be about 35dB less than the total noise power, meaning that even when the effective noise has a ratio as low as 40dB at the measurement level individual harmonics of less than 60dB of the fundamental (0·1%) can easily be measured without significant error due to the wideband noise.

## INTERMODULATION DISTORTION

Like harmonic distortion, IMD is another effect of non-linearity but when more than one input frequency is applied. Thus, while the measurements of THD requires a single input sinewave signal of

## Amplifier Tests

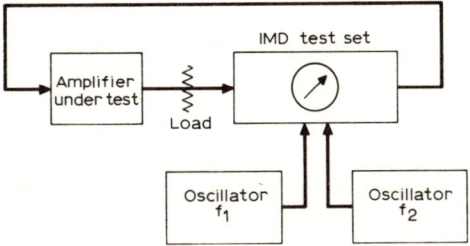

Fig. 2.10  Instrument setup for IMD measurement.

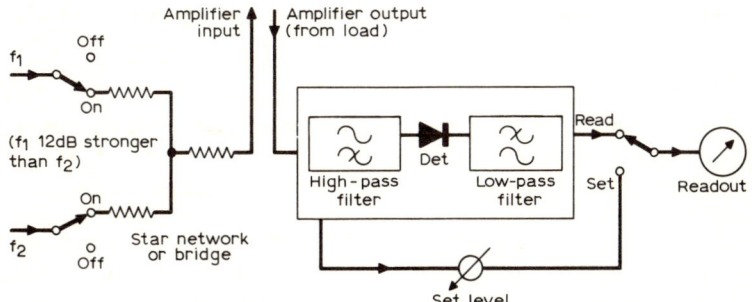

Fig. 2.11  Basic elements of r.m.s. sum IMD instrument.

high purity, the measurement of IMD requires more than one input signal.

Two signals $f1$ and $f2$ at frequencies of about 100Hz and 5kHz respectively are commonly employed in the amplitude ratio of 4:1. Some instruments include the two signal sources, while others require external oscillators, the instrument setup then being as shown in Fig. 2.10.

The two signals are fed to the IMD test set and therein combined by a network (i.e. matrix) which produces no intermodulation. The combined signal then goes to the input of the amplifier under test. The amplifier is loaded and the combined signal at the output is fed to the input of the IMD test set.

Non-linearity in the amplifier under test causes signal $f2$ to be modulated by signal $f1$ and its harmonics (Chapter 1), and on passing through the instrument signal $f1$ is removed by means of a high-pass filter, so that only the modulated $f2$ signal remains.

This is rectified (demodulated, so to speak) and directed through a low-pass filter, the residue then being the low-frequency components previously modulating the high-frequency signal $f2$. These are analysed by the instrument, and the readout is effectively the percentage modulation of $f2$ by $f1$.

Signal $f1$ will show distortion at the bass end, due to transformer saturation, etc., while $f2$ will be more in line with average operation. A switched attenuator is incorporated so that a datum can be initially established and the IMD read directly from the meter scale as a dB ratio or percentage. The basic elements of this kind of IMD test set are revealed in Fig. 2.11.

### Other IMD Test Methods

The set-up as shown in Fig. 2.11 is based on the r.m.s. sum method and is that most commonly adopted. There are other methods, including the difference-frequency method (where $f1$ and $f2$ have the same amplitude and a wave analyser may be used to measure the difference-frequency), the individual sideband method, the peak sum method and an oscilloscope method.

Unfortunately, these various methods and the differences in testing procedures tend to make interpretation of the results somewhat confusing when used as a basis for comparing the non-linearity of one amplifier with another. This does not happen with THD measurements, for the results are closely repeatable.

Nevertheless, IMD measurement does not call for signal sources of such high sinewave purity as the source required for accurate THD measurement, and it is held by some workers that a IMD readout correlates more closely to how an amplifier will 'sound' than that of THD, especially when the distortion is arising essentially from non-linearity at the middle of the Class B transfer characteristic.

However, this is debatable in my judgement, and I still prefer to compare the distortion performance of amplifiers mostly on the basis of an accurate THD readout *supplemented with waveform analysis* (using an oscilloscope). This, I feel, tells a great deal about the overall non-linearity of a hi-fi amplifier and its Class B performance.

Attempts have been made over the years to discover a simple relationship between harmonic and intermodulation distortion, but without success.

### Causes of Distortion

Distortion in low-level Class A amplifiers is often caused by incorrect transistor biasing. This encourages second harmonic distortion, and in the best of amplifiers the biasing should be optimised

## Amplifier Tests

such that the distortion remains below 0·1% even when the amplifier is driven to a dB or so below waveform clipping level.

Thus, if the distortion tends to rise towards, say 0·5% when the amplifier is driven to a level corresponding to 4dB or so below waveform clipping, then attention should be directed to the biasing resistors, which could have altered in value somewhat. It has been known for changes in value up to 20% to occur and to contribute significantly to the overall distortion even at moderately low levels of drive.

If the distortion is excessive, of course, one would usually make an immediate check of the transistor operating potentials, which is the normal servicing technique adopted to discover faulty or changed-value components.

Excessive distortion in Class B power amplifiers should, again, lead initially to a check of the operating potentials and the transistor emitter/collector currents.

Usually, however, the audio technician is concerned with locating the source of relatively small amounts of distortion. Excessive distortion, being caused almost certainly by incorrect potentials or faulty components, is generally less difficult to trace and remedy.

**Reducing Distortion**

Obviously, when analysing an amplifier for small distortion a distortion measuring test set is an essential requirement. From the normal servicing point of view there is little incentive to reduce the distortion below the value specified for the design.

This is not to say, however, that the distortion could not be further reduced by careful tailoring of the design, but since this is a design exercise it might fall outside the scope of the audio technician's brief.

Nevertheless, improved performance is sometimes possible merely by optimising the settings of the power amplifier presets. It is not unknown for an amplifier to reach its destination with both the mid-point voltage preset and the quiescent current preset maladjusted. Transit vibration could have been responsible!

The first action, therefore, is to set the mid-point voltage by adjusting the appropriate preset (e.g. VR1 in Fig. 2. ) so that the clipping at full drive is symmetrical. Asymmetry could be as bad as indicated by the oscillogram in Fig. 2.3. The plan is to get it as good as shown by Fig. 2.2, and this will require the drive to be reduced until very slight clipping is present on each half cycle peak.

After this, the quiescent current preset should be adjusted. Now, one has to be careful where the current meter is inserted to avoid the introduction of meter resistance. Where possible, the best plan is to measure the voltage across the emitter resistor of the output pair with a d.c. millivoltmeter. An ideal resistor in Fig. 2.5, for example, is the 1 ohm component in series with TR7 emitter. Since this has a value of 1 ohm, each millivolt measured will correspond to 1mA of current ($I = V/R$, where $I$ is the current in mA and $V$ the voltage in mV).

The service manual should give the best quiescent current (about 25 mA for the circuit in Fig. 2.5), so in this case VR2 would be adjusted for 25mV across the resistor mentioned. Connection here is convenient because one side of the resistor is at chassis potential.

**Warm-up Time and Temperature**

The increasing temperature of the transistors can affect the quiescent current, so it is desirable to allow a fair 'warm-up' time before finalising the adjustment. Indeed, the same applies to all critical adjustments and measurements. I usually let the equipment run at 3dB below full power for a period of at least one hour before making any adjustments or measurements. BS3860:1965 recommends that transistor equipment shall be in use for at least *two hours* before measurements are made.

It is also recommended that the ambient temperature shall lie in the range 20–25 deg. C, and for transistor equipment shall be within 2 deg. C of the maximum stated by the manufacturer, otherwise at 45 $\pm$ 2 deg. C. For transistor equipment, BS3860:1965 also states that during the 'warm-up' period the amplifier output shall be adjusted so that the maximum dissipation of energy occurs at a signal frequency of 1kHz.

After making the mid-voltage point and quiescent current adjustments the distortion should be measured and compared with that specified. Because the distortion in Class B amplifiers sometimes tends to rise as the power is reduced, particularly when the quiescent current deviates from the optimum, distortion checks should also

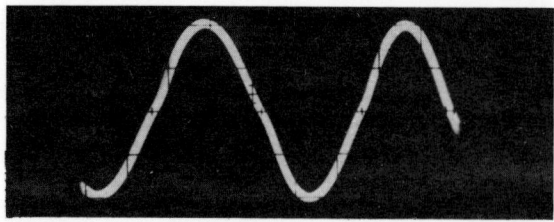

Fig. 2.12   Mild crossover distortion on a high-frequency sinewave. This is seen much more dramatically on the THD residual.

## Amplifier Tests

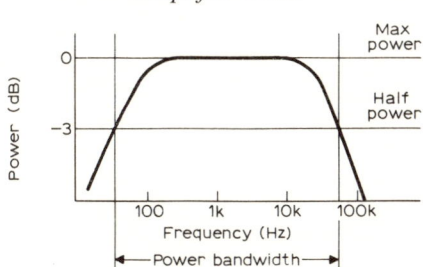

Fig. 2.13 Half power bandwidth.

be performed at powers below the rated power, preferably down to noise.

If the distortion tends to rise significantly at low powers, then the quiescent current may fail to match the requirements for the least distortion over the dynamic range. It could be insufficient. For optimum results, therefore, the quiescent current can be adjusted for the least distortion, rather than relying absolutely on the recommended current.

When adjusting in this manner it is desirable to observe the distortion/noise residual on an oscilloscope, for, as already mentioned, this can give a clue as to the degree of crossover distortion present. It is also sometimes possible to observe mild crossover distortion on the sinewave across the load, as shown in Fig. 2.12.

High-frequency distortion of this kind can be associated with hole storage effects of the output transistors, especially when the $f_T$ is relatively low.

Low-frequency distortion can be introduced by poor or insufficiently high value electrolytic coupling capacitors and by the dissipation change of the power amplifier input transistor during a voltage cycle, this latter producing a non-linear change in the input transistor and hence a rise of distortion at low frequencies.

Distortion can also arise in the tone control circuits, and a maximum THD of 0·2% has been observed even in some of the hi-fi designs. This is essentially a function of the circuit design, and various schemes are currently being evolved to reduce it.

## POWER BANDWIDTH

This is sometimes defined as shown in Fig. 2.13. Here the maximum power is that at 1kHz where the distortion is at the specified value. The curve implies, therefore, that the power at the specified distortion remains constant from about 200Hz to 10kHz, and that at lower and higher frequencies the power needs to be reduced to maintain the specified distortion.

On this basis, the power response is that between the −3dB points. An alternative (but less meaningful) measurement relates to the bandwidth between the −3dB points when a constant amplitude input signal, adjusted to provide the rated output and specified distortion at 1kHz, is changed in frequency over the spectrum. The terminal frequencies at which the power falls −3dB from that at 1kHz are noted on an audio voltmeter connected across the load. This fails to take into account the distortion at frequencies other than 1kHz.

Another presentation is related to a power change of ±1dB (or other value) over the measured spectrum when a constant amplitude input signal is applied, which, of course, differs from the half power bandwidth.

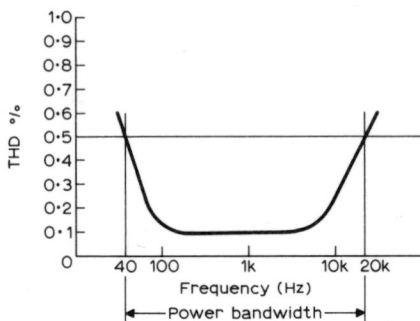

Fig. 2.14    Power bandwidth referred to THD.

The most convenient measurement is that which shows a response over a range of frequencies when the input signal amplitude is fixed or corrected at each frequency test point, while the most meaningful is that which shows a response related to THD, as illustrated in Fig. 2.14. This can be rearranged to show a power contour at the specified value of THD.

The curve in Fig. 2.14 implies that the THD remains at about 0·1% from about 100Hz to 8kHz, then rising fairly quickly at the extremes. Such a curve would be plotted at or near to the rated output power.

The instrument setup would be similar to that given in Fig. 2.1 and the audio voltmeter would have a dB scale with maximum expansion at the established datum, while the audio generator would either have a very constant output over the spectrum or

## Amplifier Tests

some means of measuring the signal level at the various test frequencies.

### FREQUENCY RESPONSE

The frequency response measurement differs from the power response measurement in that the amplifier is operated at relatively low nominal power during the former. A common level is −10dB of the rated power or less.

The basic test setup is given in Fig. 2.15. Here the audio voltmeter is used essentially to provide a readout datum. The amplifier is adjusted as required for the particular frequency response test and the audio oscillator tuned initially to 1kHz. The switched attenuator is next adjusted to provide the required level of signal at the load, while the audio voltmeter is set to a convenient datum reading.

Assuming that the audio oscillator has a constant output over the spectrum (within $\pm 0.5$dB) or that some means of measuring the signal at the various test frequencies is available, the oscillator is tuned in steps over the frequency range of interest, and the change of input required to maintain the originally established datum reading is measured in decibels.

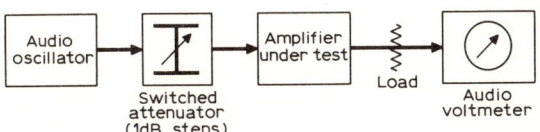

Fig. 2.15   Setup for measuring frequency response.

This measurement is performed by the switched attenuator which should be calibrated in 1dB steps, at least, over a suitable range (attenuators calibrated from 0 to 100dB are useful for tests of this nature). Some top flight audio oscillators are equipped with accurate attenuators, working in conjunction with moving-coil or digital readouts.

**Alternative Test**

When the output from the audio oscillator is not constant over the spectrum and when there is no simple means of measuring it at each test frequency, then the arrangement shown in Fig. 2.16 may be possible. Here the amplifier is driven at 1kHz to the test power with S1 at position B, and the audio millivoltmeter adjusted to give a convenient datum reading (somewhere towards three-quarters of f.s.d.).

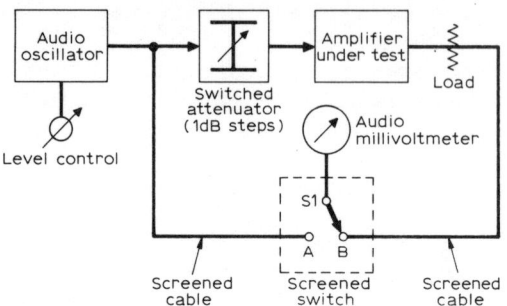

Fig. 2.16 Alternative method for measuring frequency response. (See text.)

The switched attenuator is then adjusted significantly to reduce the signal fed to the amplifier and, with S1 at position A, the level controls of the audio oscillator adjusted to give the same datum reading as before on the audio millivoltmeter without changing its formerly established setting. S1 is next changed over to position B and the attenuator adjusted to give the same datum reading as before.

Now, under this condition it should be possible to operate S1 over A and B without the audio millivoltmeter reading changing from the 1kHz datum. This is because the effective gain due to the amplifier is just being balanced by the loss due to the attenuator.

Clearly, then, any change in the effective gain of the amplifier at frequencies other than 1kHz will need to be balanced by adjusting the attenuator up or down (note that absolute gain measurements must take account of impedances). At each test frequency, therefore, the datum is noted with S1 at position A. The switch is then turned to position B and the amount of attenuator adjustment required to restore the datum reading is a direct measure (in decibels) of the frequency response deviation relative to 1kHz.

Obviously, deviations from the 1kHz zero reference established on the attenuator will have to be translated in terms of amplifier gain or loss at the other various test frequencies. For example, when the switching increases the attenuation the amplifier gain rises, and vice versa.

This is one of the most accurate ways of testing frequency response, since the measurement accuracy is related directly to the

## Amplifier Tests

accuracy of the switched attenuator and not to the accuracy of the audio millivoltmeter or audio oscillator. However, care must be exercised with regard to the input matching (oscillator to attenuator and attenuator to amplifier), and both the changeover switch and its input and output circuits should be screened.

For the sake of simplicity the 'earthy' side of the test circuit is not shown. Hum loop conditions can develop with tests of this nature, so there should only be a single 'earthing' point. This sort of trouble is diminished, however, by the use of battery powered test equipment.

**Tone Control Test**

The setup shown in Fig. 2.15 is generally used to measure the characteristics of the tone controls and the various filters, including that for 'loudness' when employed. The same applies to the equalised inputs for magnetic pickup and (sometimes) tape head.

Fig. 2.17 approximates the RIAA recording characteristic in broken line and the replay characteristic (the reciprocal of that for recording) in full line. The direct response when the test signal is applied to the magnetic pickup input should thus be close to the full line curve in Fig. 2.17. (See also Table 4.1, page 120.)

It is noteworthy that the Sugden Si453 audio oscillator (see Chapter 1) incorporates a filter tailored to the RIAA recording characteristic, so that when this is switched in the output signal

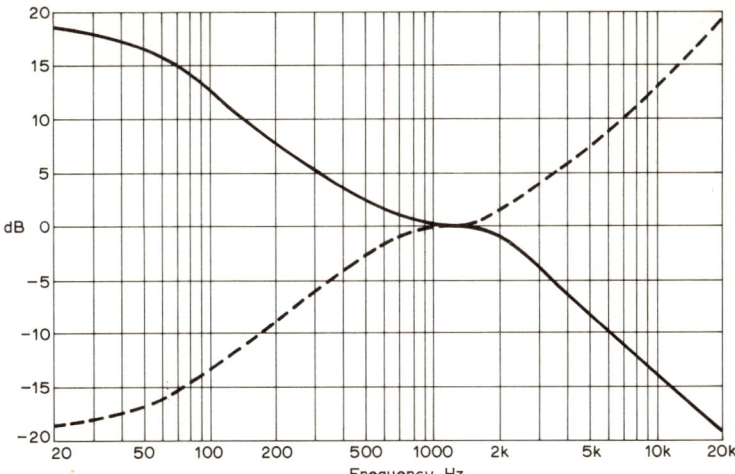

Fig. 2.17 RIAA replay characteristic full line and recording characteristic broken line.

gradually increases with frequency. This makes it possible to plot a curve showing how the magnetic pickup response deviates from 'flat' when receiving an RIAA signal (such as delivered by a magnetic pickup).

**Response Troubles**

In valve amplifiers, a poor power response is not uncommonly caused by a deficient or defective output transformer. Transistor amplifiers cannot suffer in this way because in general there is no output transformer.

Falling bass is commonly caused by coupling electrolytics of insufficiently high value. This can also encourage low-frequency distortion at an output well below the 1kHz rated power. The negative feedback is also sometimes modified by this cause, though both effects are revealed by comparing the 1kHz output resistance with that at 40Hz.

High-frequency response of transistor amplifiers is sometimes embarrassingly extended rather than inhibited, and designers then find it necessary to include a low-pass filter deliberately to roll-off the top end to prevent it from penetrating too deeply into the r.f. spectrum.

We have already seen (Chapter 1) that the h.f. response needs to extend beyond the audio spectrum for good transient performance, but there is a limit! The latest Radford amplifier, for example, has its $-3$dB point set at 160kHz by a simple RC filter. If the filter is removed the power continues to 500kHz!

Conversely, some of the early transistor amplifiers had trouble in maintaining the response to 20kHz or so. This was due to the low $f_T$ of early germanium transistors.

When tests prove that the frequency response—as distinct from the power response—is wanting, attention should be directed to the coupling electrolytics at the bass end and to the transistors and their collector load resistors at the treble end.

Excessive gain—indicated by abnormally high sensitivity—coupled with poor frequency and power response could be caused by change in value or failure of a component in the a.c. feedback loop.

## DAMPING FACTOR

The damping factor is the ratio of the specified load resistance to the output resistance of the amplifier, which is assumed to be

## Amplifier Tests

non-reactive. One way of measuring the damping factor is shown in Fig. 2.18.

Here the amplifier is receiving sinewave signal from an audio oscillator, the level of which is adjusted to produce one-quarter of the amplifier's rated power in the load when switch S1 is at position B. At this power the voltage across the load is accurately measured by the audio voltmeter to give $V_r$.

Without changing the conditions, S1 is turned to position A to remove the load which, normally, will cause the audio voltmeter reading to rise slightly. This should be carefully noted at $V_0$. The damping factor is then equal to $V_r/V_0 - V_r$.

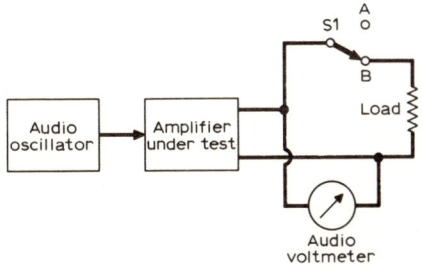

Fig. 2.18 Instrument setup suitable for measuring damping factor (see text).

The load should correspond to the value specified for the amplifier, and to conform to BS3860:1965 the test frequency should be 50Hz. Thus, if $V_r$ is, say, 4V and $V_0$ is 4·2V the damping factor at the particular load value used is 20. A load of 8Ω would then mean that the output resistance of the amplifier is 0·4Ω.

A difficulty of this test lies in the accurate measurement of $V_0 - V_r$. This is simplified by first measuring $V_r$ to get the correct power in the load and then using a valve voltmeter or electronic test meter on a low range which has provision for 'balancing out' (to give zero indication) an applied voltage.

It is then possible to get zero reading on $V_r$ and measure $V_0$ much more accurately. There are, of course, other schemes for getting the same net result. There are also other methods of measurement, one based on the switching between two loads at 1Ω and 8 or 16Ω.

The amplifier output resistance and hence the damping factor is a function of the negative feedback, and some of the latest amplifiers with 60dB or so of negative feedback provide an output resistance as low as 0·1Ω, which corresponds to a damping factor of 80 at 8Ω.

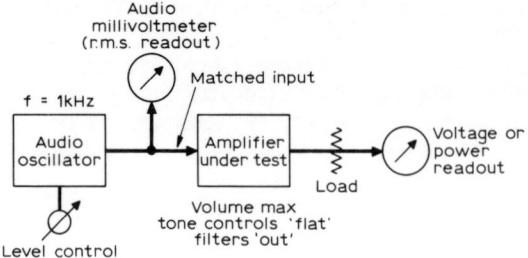

Fig. 2.19  Instrument setup for testing input voltage sensitivity.

## Aspects of Damping

A low output resistance and hence high damping factor is desirable for keeping the connected loudspeaker under electromagnetic control, but it should be appreciated that the resistance of the speech coil and connecting leads appears in series with the output resistance, so from the actual loudspeaker damping point of view the resistance is greater than the amplifier's output resistance.

Nevertheless, reasonable damping is desirable at least at the bass resonance of the loudspeaker system to minimise overshoot, etc. The specification rarely gives the frequency at which the damping factor was measured, and this could be at 1kHz, in spite of the 50Hz B.S.I. recommendation.

The damping factor assumes a resistive source, but in reality impedance components are present due to the capacitive reactance ($X_c$) of the loudspeaker coupling capacitor and sometimes the series inductance which might also be present (the resistance of which, incidentally, adds in series with the output resistance). This means, then, that the damping factor can alter with frequency, and it is not uncommon for it to fall with reducing frequency as the $X_c$ of the coupling capacitor rises.

However, some designers achieve a virtually constant damping factor with frequency, and to some extent this appears to be influenced by whether the feedback is derived before or after the loudspeaker coupling capacitor. It is also likely that the *feedback* is affected by reactance in some designs.

The damping factor is infinite when the output signal voltage remains at the same value with or without a load. It has been known for the signal voltage to *rise* when the load is connected, indicating a negative output resistance and that the amplifier is possibly on the verge of instability!

## SENSITIVITY VOLTAGE

The sensitivity voltage is a measure of the e.m.f. applied to the selected input, in series with the stated source resistance, to yield the rated output power or voltage. The test is commonly made at 1kHz, using the instrument setup shown in Fig. 2.19.

The level of the audio signal applied to the amplifier is increased to give the rated power or voltage readout, and the r.m.s. voltage is read from the input audio millivoltmeter.

## MAXIMUM INPUT VOLTAGE

When an input is connected direct to a preamplifier—such as the magnetic pickup equalised preamplifier—without a volume or level control, it is desirable to know the maximum input that the preamplifier will accept before it clips the peaks of a sinewave or introduces more than a specified value of distortion.

Fig. 2.20 shows the instrument setup. To avoid errors due to the setting of the amplifier controls, etc., the output signal can be examined at the socket delivering signal for tape recording. The audio oscillator is connected to the selected input (e.g. magnetic pickup, microphone, tape head, etc.) and its r.m.s. value measured, as in Fig. 2.19, on an audio millivoltmeter.

If the overload is to be referred to waveform clipping, then the signal is monitored on an oscilloscope and the input level increased until the tips of the sinewave just start to clip. The input is then reduced very slightly, and the overload voltage is taken from the input millivoltmeter. The test is commonly made at 1kHz.

A distortion test set is necessary to refer the overload voltage to a specified value of distortion.

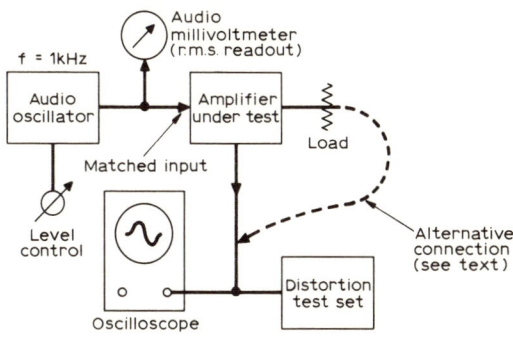

Fig. 2.20  Instrument setup for testing input amplifier overload voltage.

With non-integrated amplifiers the signal can be obtained from the control section output socket, assuming that due attention is given to the setting of the controls (tone controls 'flat', filters 'out' and the volume control turned well down to avoid overloading of the stages after the preamplifier).

An alternative arrangement is shown by the broken line connection in Fig. 2.20. Here the signal is monitored and its distortion measured at the output of the power amplifier. For this to be meaningful the controls should be set as mentioned above and the volume control must be turned down very low to prevent overloading of the power amplifier.

This scheme has the disadvantage that the tone controls, primary amplifiers and power amplifier contribute to the distortion when the overload is referred to this.

**Aspects of Input Overload**

Of most importance is the overload capability of the magnetic pickup equalised preamplifier, for if this tends to overload too early excessive distortion can result from discs carrying high velocity modulation. The overload is often given as a dB ratio relative to the input sensitivity, and a ratio much less than 25dB (about 36mV relative to an input sensitivity of 2mV, for example) can be too close to overload for comfort.

It is normal for an RIAA preamplifier to reduce in overload margin as the signal frequency falls. This is due to the action of the equalising and is common to all circuits of this kind.

However, few amplifiers have a volume control stage prior to the RIAA preamplifier, and with these it is virtually impossible to overload the RIAA preamplifier under normal conditions of operation.

Generally, though, the magnetic pickup connects direct to the preamplifier for reasons of optimising signal/noise performance.

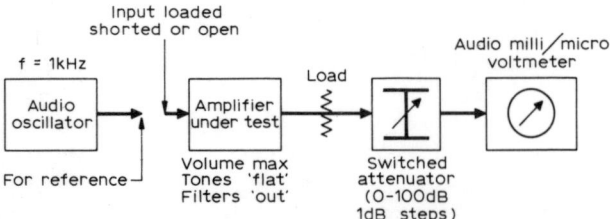

Fig. 2.21    Instrument setup for S/N testing.

and this is the type of amplifier which is most critical from the overload aspect.

Poor overload performance can mean that the designer has veered more in favour of enhanced signal/noise performance than overload performance. The two factors tend to conflict in design.

## HUM AND NOISE

The level of signal due to residual hum and noise across the output load of an amplifier is commonly referred to the rated output as a dB ratio, and is sometimes called the signal-to-noise (S/N) ratio. Thus, an amplifier with a rated output of 10W would have an S/N ratio of 40dB when the power due to the hum and noise is 10,000 times below 10W, or 1mW.

In practice, the ratio is generally measured by referring the *voltage* across a specified value of load resistance due to the hum and noise to the *voltage* of the signal across the same load when the amplifier is driven to its rated output with a 1kHz sinewave. For example, an amplifier which, at its rated output, produces 12V r.m.s. across 8Ω (corresponding to about 18W) when driven by a 1kHz sinewave would have an S/N ratio of 40dB when the hum and noise across the same load corresponds to 120mV.

### S + N/N Ratio

When tested like this, however. it is the S + N/N ratio that is measured. Not the true S/N ratio, but the error is insignificant when the S/N ratio is not less than about 20dB. For smaller ratios than this the measurement should take into account the noise on the signal to which the noise alone is being referred; but this rarely applies to the type of equipment which passes through our hands.

It is noteworthy that the same principle applies to the measurement of THD, but since the distortion is so small compared with the signal, correction is rarely necessary.

The S/N test setup is given in Fig. 2.21. The audio oscillator is connected to the input under test and the amplifier switched to select this and adjusted as shown. The attenuator is switched to maximum value and the input signal and audio milli/micro voltmeter adjusted to provide a suitable readout datum at the rated power into the load.

The generator is disconnected and the selected input shorted across. The idea, then, is to switch down the attenuation until the originally established datum is obtained, the S/N ratio in decibels being read direct from the attenuator which, of course, should be calibrated.

The test can be repeated, if required, with the selected input open-circuit or loaded with the specified value of resistance. In the latter case the load should be screened to avoid electrostatically or magnetically induced hum.

This measurement is, of course, the noise *plus* the hum, so in the strict sense it is not the S/N ratio. Nevertheless, the S/N item in most specifications relates to this type of measurement unless otherwise stated. The residual hum and noise can be measured with the volume control turned right down, and this is sometimes given in terms of voltage across the load.

It will be appreciated that the measured noise is that over the full effective bandwidth of the amplifier. In some cases the use of filters will be necessary to remove the hum, noise or other interference when it is required to exclude its effect from the measurement. More information on this is given in Chapter 3.

Moreover, the ratio may be weighted as explained in Chapter 1, page 8 relative to the weighting curve in Fig. 1.2.

## EQUIVALENT NOISE INPUT VOLTAGE

Some specifications give the hum and noise (or just the noise with the hum filtered out) as an equivalent noise voltage in series with the stated source resistance at the input. The definition of the equivalent noise input voltage is the input voltage at 1kHz, in series with the specified source resistance, that would produce an output at the specified load equal to that produced by the hum and noise or noise alone when the hum is filtered out.

This can be expressed as a dB ratio by referring the equivalent noise input voltage to the sensitivity voltage at the particular input at 1kHz, when the input is loaded by the specified source resistance.

### Causes of Hum and Noise

Excessive hum would indicate a shortcoming in the rail voltage smoothing circuits when present with the selected input shorted. Should the hum level rise when the input is properly terminated with a screened load equal to the specified source resistance, then hum pickup on the input circuits would be a strong possibility.

It has not been unknown for magnetic radiation from the mains transformer to cause a hum-loop condition in conjunction with the

*Amplifier Tests* 67

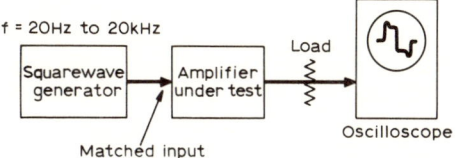

Fig. 2.22   Instrument setup for squarewave testing.

'live' power amplifier output cable. The nature of the hum can be observed on an oscilloscope connected to the output socket of the audio micro/milli voltmeter. 100Hz hum indicates poor smoothing and 50Hz hum direct induction or pickup from the mains supply circuits or associated components. The 50Hz waveform might carry third-harmonic distortion, while the 100Hz waveform is reasonably 'pure'.

Excessive noise often arises from a 'noisy' resistor or valve in a high-gain early stage. In solidstate equipment it is possible for a preamplifier transistor or transistors to rise in noise output, and this applies especially to the transistors used in the RIAA magnetic pickup preamplifiers.

It is easily possible to compare the noise performance of one channel with that of the other in stereo amplifiers as a means of proving whether the noise is excessive for the particular design. Both channels rarely go down together in this respect, though it is possible for this to happen.

Below the rated output, the hum and noise of a good hi-fi amplifier will not be worse than 70dB at all the inputs except magnetic pickup, tape head and microphone, where it should not be worse than 60dB. The residual hum and noise (with the volume control turned right down) should not be worse than 80dB. It is usually better than this—better than 100dB with some top grade designs.

'Noise' caused by electrical interference, radio breakthrough and the like is considered in Chapter 5.

## SQUAREWAVE TESTS

A squarewave input when monitored at the output by an oscilloscope provides a good impression of the amplifier's rise/fall time, response characteristics and transient performance, as already noted in Chapter 1.

The instrument setup is given in Fig. 2.22. The squarewave signal is applied to the auxiliary or radio input (or to the main input of a power amplifier) and the output is monitored across the load resistor. The effective Y bandwidth of the oscilloscope should be several times greater than that of the amplifier under test.

The oscilloscope should also have sweep velocity calibration in conjunction with a screen graticule scribed in cm squares with mm markers. The sweep velocity is generally given in terms of time per centimetre of sweep. Thus, a sweep of 1μS/cm is the time it takes the spot to move over 1cm of screen.

The 50Hz mains supply cycle, for example, occupies a time period of 20mS (time in mS equals 1,000/frequency), which means that when the sweep is adjusted to 20mS/cm one complete cycle of 50Hz signal will occupy 1cm of screen. Similarly, a sweep of 1μS/cm will display a 1MHz full cycle over 1cm of screen.

Provided that the rise time of the oscilloscope's Y amplifier is several times greater than that of the amplifier under test, the latter can be measured when the rise time of the signal itself is greater than that of the amplifier.

The plan is to get the oscilloscope to trigger on the rising side of the squarewave as shown in Fig. 2.23(a). The rise time is that period of time shown by the graticule corresponding to the time taken by the wave to rise between 10 and 90% of its full amplitude. This is

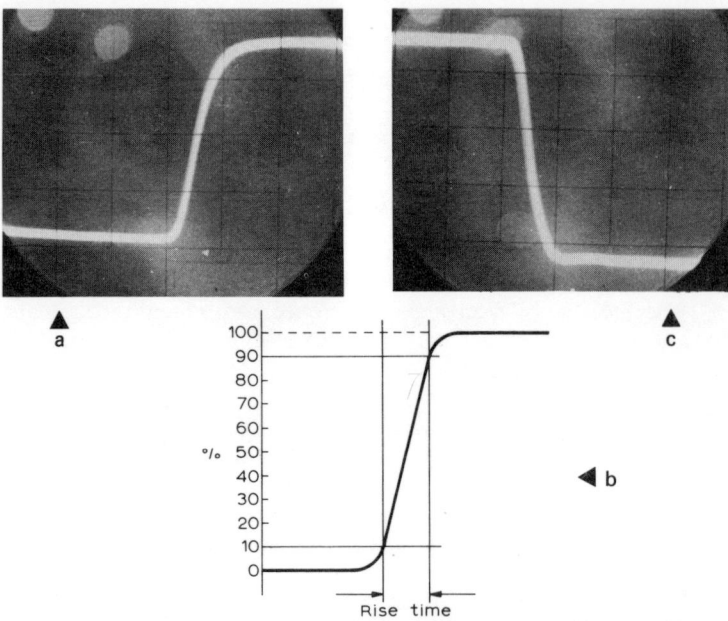

Fig. 2.23  Rise time oscillogram (a) and definition (b). Fall time oscillogram (c).

## Amplifier Tests

made clear in Fig. 2.23(b). The falling side of the squarewave can also be caused to trigger the oscilloscope as shown in Fig. 2.23(c), and the fall time can then be measured in a similar manner.

### Rise Time

The rise time is a function of the amplifier's −3dB bandwidth which is approximately equal to 400/rise time in µS, the frequency then being in kHz. Thus, an amplifier with a 4µS rise time would extend to about 100kHz in response down to −3dB. To check this, therefore, the rise time of the oscilloscope's Y amplifier and the rise time of the squarewave should be significantly less than 4µS, but this is not a problem for reasonable oscilloscopes have a Y amplifier rise time of 0·1µS and squarewave generators of the same order.

As is well known, a squarewave is composed of a sinewave equal to the fundamental or repetition frequency of the signal and a series of odd-numbered sinewave harmonics of specific amplitude and phase relationships. Thus, if the squarewave is to pass through the amplifier without distortion the phase and amplitude relationships of the harmonics must be maintained.

The majority of audio amplifiers handle a squarewave at 1kHz without changing its character when the tone controls are 'flat' and the filters switched out. Indeed, a squarewave at 1kHz is useful for checking the 'flat' settings of the tone controls.

A dual-trace oscilloscope makes it possible to view both the input and the output waveforms simultaneously, and when the tone controls (bass and treble) are set so that the output waveform is a virtual replica of that at the input, then one can be assured that the controls are essentially at neutral.

### Typical Waveforms

The series of squarewave oscillograms in Fig. 2.24 gives some idea of the results that can be expected. (a) is a fair 1kHz display with the tone controls flat and with the amplifier extending to about 30kHz. (b) shows what happens when the treble response has an early roll-off or when the signal frequency is increased to about 10kHz. (c) is the display to be expected at 10kHz when the treble starts to roll-off at about 20kHz.

This sort of display (and that at (b)) would also be produced with the treble control giving top cut when the squarewave is at 1kHz. A waveform similar to that at (d) occurs at 10kHz when the treble has started an early fall due to an amplifier with a very poor h.f.

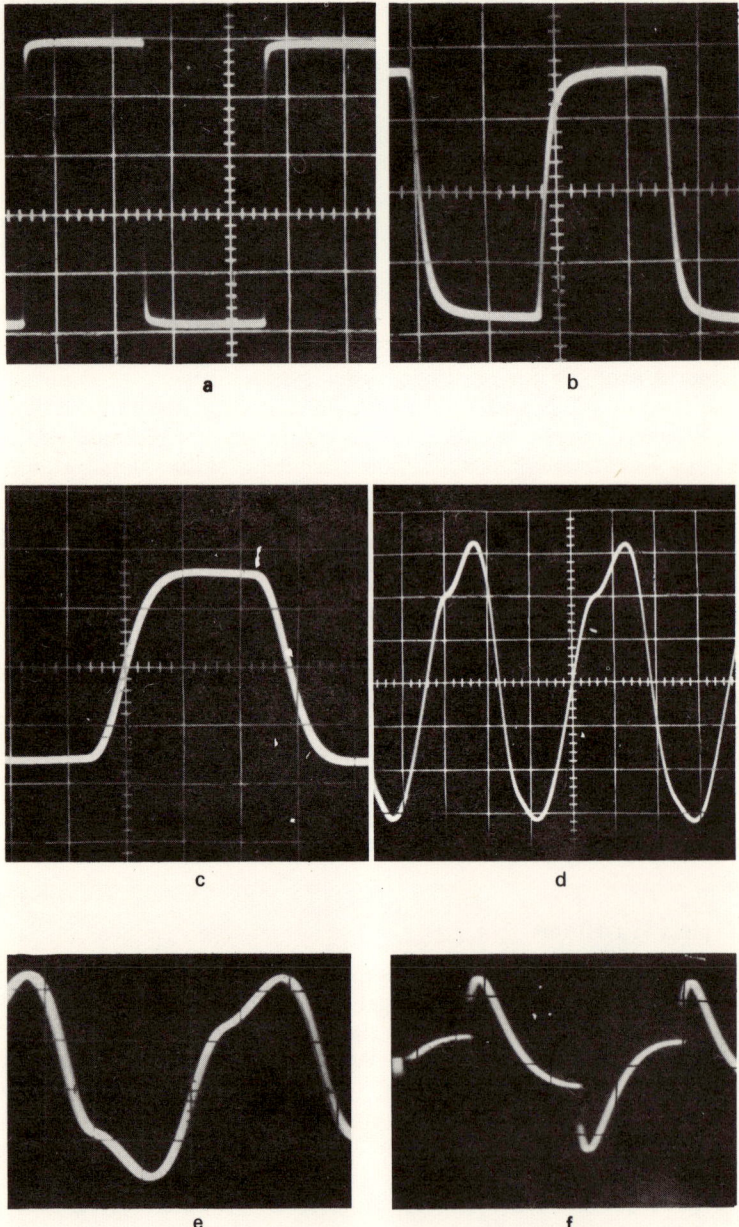

Fig. 2.24 Series of squarewave tests. See text for description.

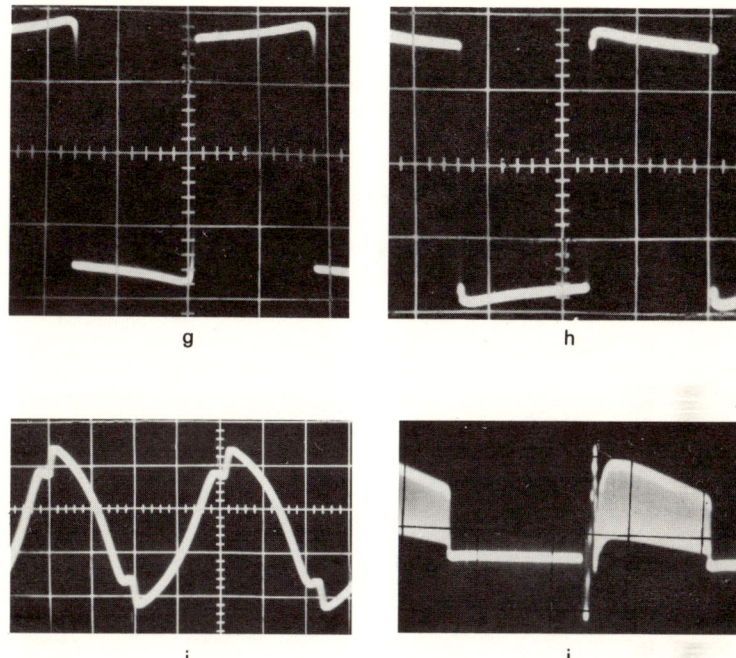

g  h

i  j

response. When there is bass lift as well as treble fall-off, the waveform appears as shown at (e).

(f) shows the effect of bass boost and possibly treble boost as well when the squarewave frequency is about 40Hz. A gradually rising top, as at (g), indicates mild bass boost when the squarewave frequency is about 40Hz, while a gradually falling top, as at (h), indicates mild treble boost when the squarewave frequency is about 10kHz.

The effect of switching in a treble filter (steep slope) is shown at (i), and finally, (j) shows what happens to the squarewave when the application of treble boost causes the amplifier to oscillate.

All these were taken across a load of pure resistance, and tests under these conditions will reveal overshoot or ringing tendencies of filters, etc. However, to simulate a loudspeaker the load should be reactive, and waveforms resulting from load reactance are given in Chapter 1 (Fig. 1.3).

Clearly, then, the squarewave represents a very revealing amplifier test and, when the amplifier is driven towards full power from such a signal, a very demanding one, too, particularly when heavy reactance is present in the load.

CHAPTER THREE

# TUNER TESTS

TUNERS NOWADAYS INCORPORATE facilities for the reception of the f.m. signals in Band II (about 88 to 108MHz) in stereo as well as in mono, the former by means of an in-built or plug-in decoder, and sometimes facilities for the reception of stations in the medium, long and short wavebands.

A multiplying breed is the tuner-amplifier or 'hi-fi receiver' as is sometimes named. This carries a stereo amplifier, so the tests relating to this will be as detailed in Chapter 2, and an f.m. stereo tuner, again with facilities in some cases for the reception of medium-frequency stations. This chapter deals with tests suitable for tuners alone and for the tuner sections of tuner-amplifiers.

### F.M. TUNER TESTS

The input to a tuner is r.f. signal and the output is a.f. signal, so basically we need to apply a signal of the former type and have facilities for measuring the latter type. With f.m. tuners, of course, the r.f. signal must be tunable over Band II, while the source must have facilities for frequency modulation, preferably from zero to at least $\pm 75$kHz deviation (100%), or if the deviation is fixed (usually at about 30%), then it should also be possible to switch the modulation off, leaving a pure Band II carrier.

While many f.m. signal generators feature an in-built audio oscillator to provide the modulation signal, certain tests require this signal to be tunable over the audio spectrum. Thus, if the in-built audio oscillator operates at a fixed frequency (usually 400Hz or 1kHz), then there should be provision for the injection of modulation signal from an external audio oscillator at a measurable deviation (modulation depth).

Moreover, for harmonic distortion tests the modulation signal should itself carry no more than 0·05% THD, while the process of frequency modulation should also be linear with minimal distortion yield.

Since THD tests in f.m. tuners are generally carried out at 1kHz, it is possible to filter the signal from a 1kHz oscillator to delete the harmonic components, thereby making a mediocre oscillator suitable for this activity. Such filtering is considered later.

For frequency response tests all the way through a tuner from aerial input to audio output, the modulation signal must, of course, be tunable from 20Hz to 20kHz, and account must be taken of the tuner de-emphasis.

Thus, there must either be a suitable pre-emphasis network between the audio oscillator and the modulator, or the level of the modulation at each spot-frequency above 1kHz must be adjusted to simulate the pre-emphasis, thereby making it possible to measure relative to a fixed readout datum.

The r.f. signal matching from the generator to the aerial input of the tuner should be accurate to avoid undue voltage standing-wave ratio, direct radiation should be minimal (below $1\mu V/m$) and the attenuators should indicate accurately the level of the signal applied to the tuner, taking into account an artificial aerial when used. The high sensitivity of modern f.m. tuners makes it necessary to measure at signal inputs below $1\mu V$ into $75\Omega$ (or $300\Omega$). (See also page 84.)

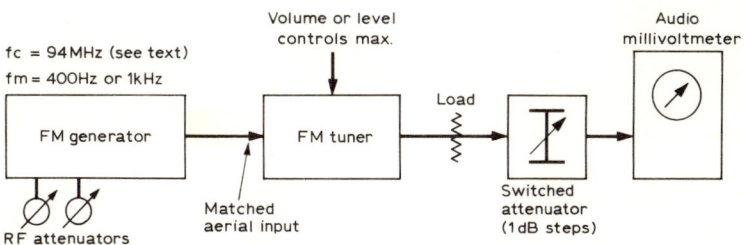

Fig. 3.1  Instrument setup for checking limiting, leading to the curves shown in Fig. 3.2. Absolute audio signal voltage is generally measured directly across the output load with the millivoltmeter (see text).

## Audio Output and Limiting

The instrument setup for measuring audio output and limiting is given in Fig. 3.1. The v.h.f. carrier frequency ($f_c$) is commonly 94MHz (or as near to this frequency as possible consistent with zero breakthrough from off-air signals) and the modulation frequency ($f_m$) 400Hz or 1kHz.

## F.M. Test Frequencies

94MHz is towards the middle of Band II, but it is sometimes desirable to make tests at other frequencies in the band, and Table 3.1 below gives the BSI groups of frequencies for tests at seven and three points and at one point. IHF test frequencies are given on page 86.

Table 3.1

| Seven | Three | One |
|---|---|---|
| MHz | MHz | MHz |
| 88 | 88 | — |
| 90 | — | — |
| 92 | — | — |
| 94 | 94 | 94 |
| 96 | — | — |
| 98 | — | — |
| 100 | 100 | — |

The load in Fig. 3.1 should correspond to that recommended by the manufacturer, and to establish a readout datum it is generally convenient initially to apply an r.f. input of 10mV with the modulation running at the selected deviation.

Any volume or level control should be turned to maximum and the receiver very accurately tuned to the input signal. Sufficient time should be allowed to elapse for maximum frequency stability, and after finally correcting the tuning the a.f.c., if embodied, should be switched on.

A convenient 0dB datum point should then be established on the audio millivoltmeter by adjusting the output attenuator (this may be incorporated in the audio millivoltmeter). The idea is then to plot a curve of audio output versus r.f. input, as shown in Fig. 3.2.

The output, of course, is relative to the established 0dB datum, and is read directly (in plus or minus values) from the output attenuator, this being adjusted up or down to retain the readout datum at each of the r.f. input levels plotted, starting from the lowest input level provided by the generator attenuators (see under *Artificial Aerials*, page 84) and continuing at levels up to and beyond limiting.

## X Axis

The scale of the curve's X axis will, of course, need to match the effective sensitivity and limiting performance of the tuner under test. The full-line curve in Fig. 3.2, for example, indicates that the

## Tuner Tests

limiting is complete at an input of about 6μV (using the unbracketed signal values on the X axis), while the broken-line curve (based on the bracketed signal values on the X axis) shows that maximum limiting occurs with an input of about 400μV.

The tuner responsible for the full-line curve, therefore, is significantly more effective so far as the limiting is concerned (and almost certainly more sensitive—see later) than the tuner responsible for the broken-line curve.

While curves like these show how the audio output varies over a range of input signals, and how well the tuner limits (i.e. arrives at maximum output, after which there is no increase in output with increasing r.f. signal input), they fail to indicate in this form the actual output voltage.

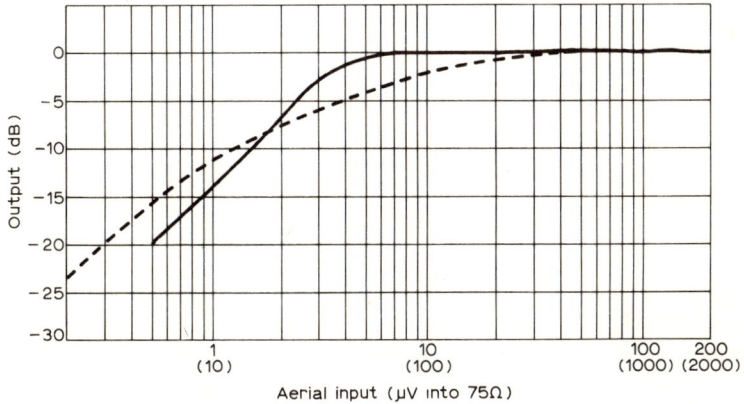

Fig. 3.2 Output versus modulated r.f. input signal, showing the point of complete limiting. The full-line curve relates to the unbracketed signal voltage values on the X axis, while the broken-line curve relates to the bracketed signal voltages.

This is best measured by using the setup in Fig. 3.1 minus the output attenuator. The audio millivoltmeter is then connected to read the audio signal actually across the load at an r.f. input giving maximum limiting, and at a specified modulation frequency and deviation.

Maximum modulation (100%) is generally regarded as ±75kHz deviation, and tests can be made either at that value or 30% of it, the latter corresponding more closely to 'average' audio output under normal operating conditions. A convenient modulation frequency is 1kHz.

## Total Harmonic Distortion

The instrument setup for measuring f.m. tuner total harmonic distortion is given in Fig. 3.3, and the distortion measuring procedure is equivalent to that explained under THD in Chapter 2.

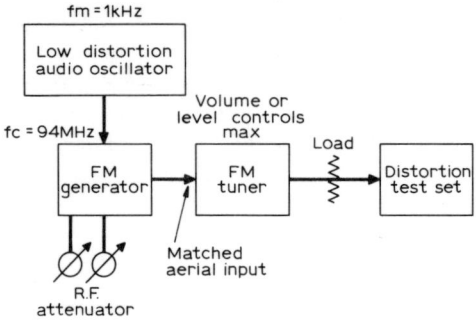

Fig. 3.3  Instrument setup for measuring f.m. tuner THD.

## AUDIO SIGNAL FILTERING

To reduce the distortion on the audio modulation signal a filter such as shown in Fig. 3.4 can be connected between the audio oscillator and the f.m. generator modulation signal input. The distortion is reduced (to about one-fifth of that on the oscillator signal direct) because at resonance the tuned circuit attenuates all harmonic components, passing with minimal attenuation the fundamental frequency.

As the tuned circuit must be of high Q-value, it is important for the oscillator to be reasonably frequency stable (a requirement for the THD test anyway), and if the action is required at a number of frequencies, then either L or C has to be variable, or a series of filters need to be constructed.

## Oscillator Distortion

Usually, however, the requirement is for the lowest oscillator distortion at the nominal 1kHz test frequency, and for this L should be 200mH and C 0·13µF. L should be air-cored, since dust-iron cores can introduce third-harmonic distortion, and it can be constructed by winding 4,200 turns of 34 s.w.g. enamelled-covered copper wire on to a former of $\frac{1}{2}$in. diameter and 1in. width.

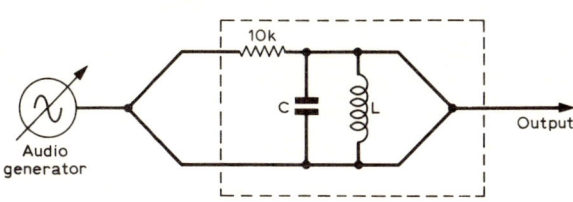

Fig. 3.4 A filter like this can be connected between the audio oscillator and the F.M. generator modulator to filter the harmonics and thus reduce the signal distortion (see text).

The f.m. generator should be adjusted to yield an r.f. signal of sufficient level to take the tuner into limiting, and the deviation should be set to $\pm 75$kKz (although, of course, it is possible to measure THD at lower modulation depths). THD at audio frequencies other than 1kHz can be tested if required, but this parameter of most f.m. tuner specifications is based on 1kHz and 100% modulation or 30%.

**Detector Non-linearity**

The distortion stems in part from non-linearity of the f.m. detector transfer characteristic and r.f./i.f. stages and in part from non-linearity of the audio preamplifier and stereo decoder following the f.m. detector, when these are active during the test. A hi-fi tuner rarely produces more than 0·5% THD under the test conditions detailed.

However, the amount of distortion will rise significantly should the tuner drift from the generator frequency. In fact, during the test it is desirable to tune for the least distortion, and it should then be noted whether this tuning point corresponds to maximum deflection on the tuning indicator if a maximum reading meter, or to 'balance' if a centre-reading meter or if it comprises a pair of tuning indicator lamps.

When tuned according to the manufacturer's instructions, by using the tuning meter or indicator, for example, not only should the distortion fall to a minimum and the a.f. output power rise to a maximum (the latter being checked when testing for limiting— Figs 3.1 and 3.2), but the noise output should also fall to a minimum and the amplitude-modulation suppression rise to a maximum, as governed by the general design of the tuner.

If these parameters fail to coincide when the tuning is correct, then the design of the tuner may be poor or the i.f. channel or f.m.-detector alignment may be in error. However, if coincidence is achieved at a tuning point slightly removed from the optimum indicated by the meter, etc., then there could be trouble in the tuning of the meter or indicator circuit itself.

## FREQUENCY RESPONSE

It is preferable to check the tuner frequency response by varying the frequency of the modulation signal, and then plotting audio frequency against output. An instrument setup suitable for this is given in Fig. 3.5. Here is shown a filter between the audio oscillator output and the f.m. generator modulator input which automatically introduces the pre-emphasis to match the de-emphasis at the tuner.

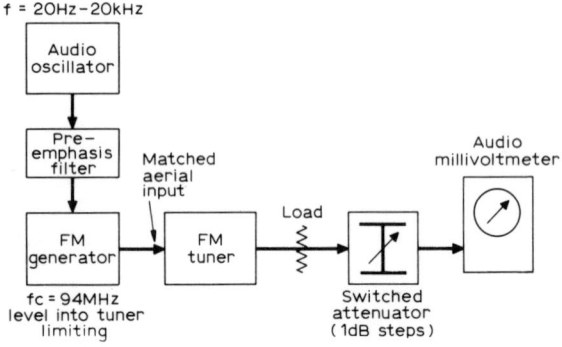

Fig. 2.5  Instrument setup for plotting f.m. tuner frequency response. A switched attenuator can be used instead of the pre-emphasis filter, as explained in the text.

The curve in Fig. 3.6 corresponds to the standard European pre-emphasis characteristic. This is based on a 50μS time-constant. The pre-emphasis in American countries is based on a 75μS time-constant, so the pre-emphasis filter used in the setup in Fig. 3.5 should correspond to the de-emphasis employed in the tuner under test. Fig. 3.7 compares 50μS pre-emphasis (full-line curve) with 75μS pre-emphasis (broken-line curve) at A, while the tuner de-emphasis characteristics are likewise compared at B.

From these curves, therefore, it will be appreciated that a tuner carrying 75μS de-emphasis will exhibit a falling treble response when the test is based on 50μS pre-emphasis, as shown in Fig. 3.8, curve B.

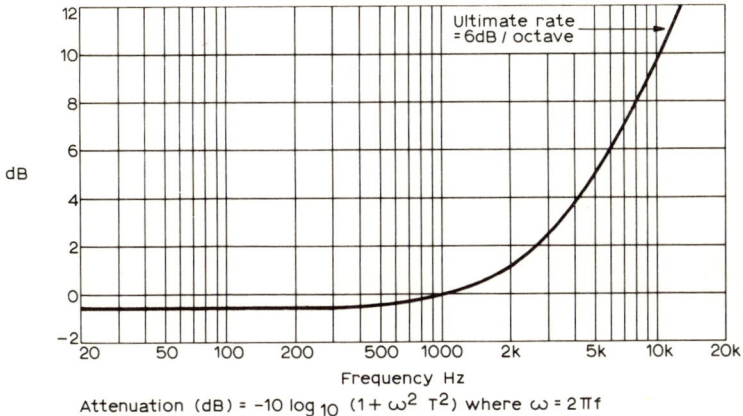

Attenuation (dB) = $-10 \log_{10} (1 + \omega^2 T^2)$ where $\omega = 2\pi f$
f = Frequency (Hz)
T = CR = time-constant (sec)
R = total effective source resistance

Fig. 3.6 Standard 50 µS pre-emphasis characteristic.

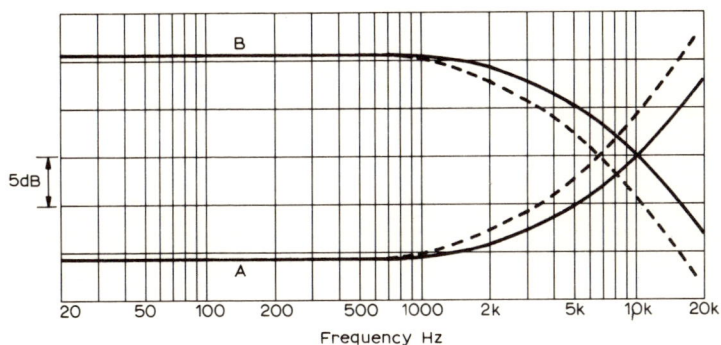

Fig. 3.7 Pre-emphasis curves A and de-emphasis curves B. Curves in full-line are 50 µS and those in broken-line 75 µS.

## Plotting Attenuation

Returning to Fig. 3.5, the plan is to establish a datum with the switched attenuator and then to plot the amount of attenuation (in decibels) that has to be added or removed at each frequency test point to retain the datum, ending up with a curve such at that at A or B in Fig. 3.8.

It is possible, of course, to plot a frequency response curve by using a switched attenuator in place of the pre-emphasis filter. In this case the 0dB datum would be established at 1kHz and the

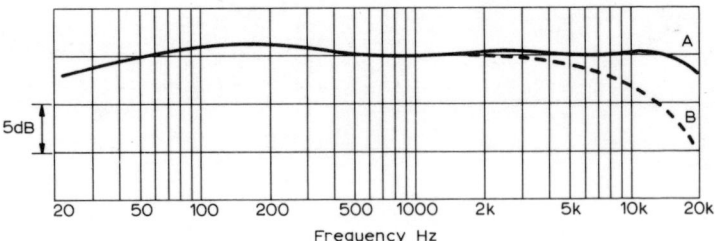

Fig. 3.8 F.M. tuner frequency response. Curve A might be obtained from a tuner designed for 50μS de-emphasis when the modulation signal is via a 50μS pre-emphasis filter, as in Fig. 3.5, while curve B would result when the tuner de-emphasis is 75μS and 50μS pre-emphasis is used.

modulation signal level adjusted by the attenuator in this circuit at each test frequency in concord with the curve (for 50μS) in Fig. 3.6. The readout would be the same as already explained.

Whichever procedure is adopted, it is important to ensure that the output from the audio oscillator is the same at each test frequency.

## Mono Tuners

A mono tuner has the de-emphasis network directly after the f.m. detector, as shown at (a) in Fig. 3.9, while a stereo tuner has a

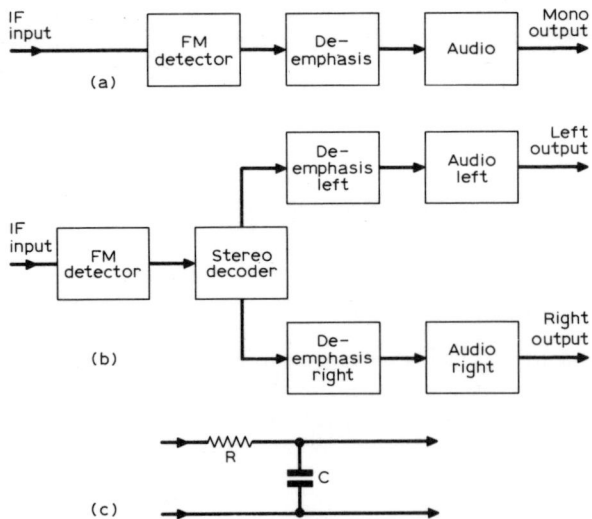

Fig. 3.9 De-emphasis. (a) in mono tuner. (b) in stereo tuner. (c) typical network.

similar network in each channel output of the decoder, as shown at (b) in Fig. 3.9. The network commonly consists of a series resistor (R) and parallel capacitor (C), as shown at (c) in Fig. 3.9, the product giving the time-constant in μS when C is in nF and R in kΩ. Thus, a 50μS time-constant would be obtained when C is 1nF and R is 50kΩ.

In practice, R is the source resistance of the circuit (or some of it is, anyway), so it is not always easy from the circuit accurately to calculate the de-emphasis time-constant. Moreover, as a means of attenuating subchannel components (e.g. the pilot tone at 19kHz and/or the reference 'switching' signal at 38kHz), a stereo tuner may contain a filter in the audio output circuit.

Since this can influence the roll-off and other characteristics of the de-emphasis time-constant, the two circuits may be designed as an integrated whole. This further complicates accurate calculation from the in-circuit component values!

## STEREO SUBCHANNEL REJECTION

A stereo tuner with such subchannel attenuation should be checked at least at 19kHz using the setup in Fig. 3.5, but without the pre-emphasis filter. The idea is to establish a 0dB datum with the modulation frequency at 1kHz at a given deviation, and then to tune accurately to 19kHz and the same deviation (at which point the stereo indicator should light on the tuner) and discover by means of the switched output attenuator the amount (in decibels) by which the response is reduced at that frequency.

The ordinary de-emphasis will account for almost 18dB, and if there is a filter this will probably add another −22dB, resulting in an overall rejection at 19kHz of about 40dB (sometimes more). If the modulator of the f.m. generator permits, a similar test should be made at 38kHz.

It is seemingly feasible to plot the characteristics of the de-emphasis and subsequent filters, when the latter are fitted in stereo tuners, by applying the audio signal across the f.m. detector load, instead of modulating it on to the v.h.f. carrier. However, this practice can lead to errors due to reactive effects, etc., in the audio coupling, so should be handled with caution.

In a stereo tuner, of course, this method of testing will also give a response check of the decoder, but for the most accurate results the decoder should be stereo-active, and this requires either a simulated stereo signal or a stereo signal generator, about which more anon.

## SIGNAL-TO-NOISE RATIO TEST

The S/N ratio of an f.m. tuner is related to the strength of the aerial input signal. A suitable test setup is given in Fig. 3.10.

The plan is to measure the ratio of the output voltage resulting from the signal to that resulting from random noise across the output load which, of course, represents the S/N ratio. In this test, hum components are filtered out, so that the resulting ratio excludes the effect of noise and hum below 300Hz and noise above 15kHz, this being handled by the 0·3–15kHz bandpass filter.

### S + N/N

The 1kHz bandpass filter in the lower branch of the test circuit ensures that the ratio is true S/N and not, as it would be without this filter, S + N/N. However, when the noise and hum components are small and the true S/N ratio is not less than about 20dB (it is usually better than this in practice at usable signal input levels), S + N/N does not differ significantly from S/N.

The test can thus be made without the 1kHz bandpass filter to get a S + N/N readout and without the 0·3–15kHz bandpass filter when the ratio is to include hum and noise below 300Hz and noise above 15kHz.

The 1kHz bandpass filter, of course, assumes that the frequency of the signal measurement in the ratio is 1kHz. For a different $f_m$ a

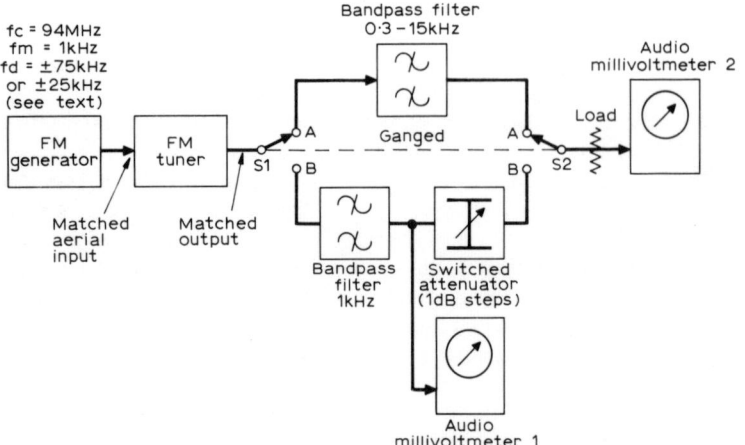

Fig. 3.10  Instrument setup for measuring S/N ratio (B.S.I.).

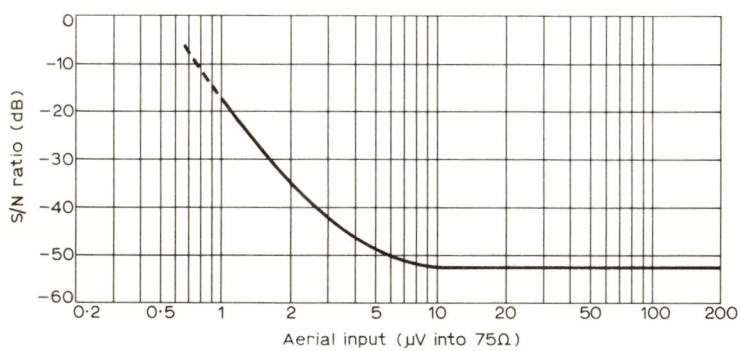

Fig. 3.11 S/N ratio curve (see text).

bandpass filter of corresponding frequency would need to be used. The matching and loading have to be arranged so that there is no significant change when S1 and S2 are operated.

Modulation depth or deviation frequency ($f_d$) can be either $\pm75$kHz or $\pm25$kHz to correspond to 100% or about 30% modulation. BSI recommend the latter and IHF the former, which results in the ratio being about 9dB greater than when referred to 30% modulation.

**Test Procedure**

S1/S2 are ganged, and with these set to position B the tuner is accurately adjusted to the selected $f_c$ with $f_m$ on, using audio millivoltmeter 1. S1/S2 are then switched to position A, and with $f_m$ switched off the sensitivity of audio millivoltmeter 2 is adjusted to give a convenient datum reading.

This meter is now reading noise, and at this stage it is desirable to check the tuning. for if this is in error the noise output will be higher than it should be. Thus, tune for the least noise and correct the readout datum accordingly.

Finally, S1/S2 are switched back to position B, $f_m$ is switched on again and the switched attenuator adjusted to give the previously established readout datum. The setting of the attenuator gives the S/N ratio directly in decibels.

A series of measurements are made like this over a range of r.f. signal levels, leading to a curve such as that shown in Fig. 3.11. This shows that at inputs above about 10μV there is no improvement in S/N ratio.

Below 10μV, however, the ratio starts to diminish, dropping

to less than −20dB at 1μV input. Such a curve indicates a relatively sensitive tuner, which would probably have a limiting curve similar to that shown by the full-line in Fig. 3.2.

## GENERATOR MATCHING

Unless a very expensive f.m. generator is employed, it is virtually impossible to obtain meaningful measurements at inputs below about 1μV. Even with expensive equipment the validity of measurements below 1μV are questionable unless one can be sure that the generator terminal impedance closely matches the input impedance of the tuner (see under artificial aerials).

Most generators have an unbalanced (i.e. for coaxial cable) output at 50–70Ω. Thus, when testing f.m. tuners with 300Ω balanced inputs a low-loss and well matched balun transformer can be used. Alternatively, it is sometimes possible to obtain a 75Ω input by connecting the braid of the generator cable to an 'earth' point on the tuner and the inner conductor to one of the 300Ω terminals; *but this is not always possible.*

### Artificial Aerials

BSI and IHF specify the use of artificial aerials for matching the generator to the tuner. These are resistive devices which reflect to the generator a pure resistive termination corresponding to the modulus of its characteristic impedance, and to the aerial terminals of the tuner a pure resistive termination corresponding to the value specified by the manufacturer.

Fig. 3.12 gives examples (BSI) of these devices. (a) and (b) are suitable for unbalanced generators and unbalanced tuners, (a) when the generator resistance is greater than the tuner resistance

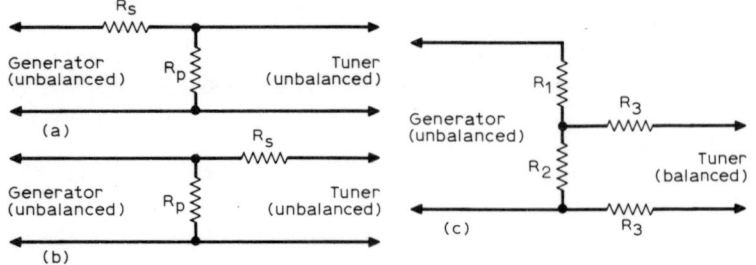

Fig. 3.12 Example artificial aerials for F.M. (B.S.I.).

## Tuner Tests

and (b) when the generator resistance is smaller than the tuner resistance.

(c) illustrates a scheme for obtaining an output for a balanced tuner from an unbalanced generator with provision for matching the signal generator. In this case, when R2 is small compared with R1 and R3, R1 + R2 serve to match the generator, while R2 + 2R3 provide the source resistance for the tuner.

Without the use of an artificial aerial or 'coupling pad' the tuner aerial input will see the internal resistance of the generator direct. A balanced IHF network is referred to on page 86.

### Aerial Input Signal Level

The signal from the generator applied to the tuner is commonly expressed as a voltage, corresponding to the *equivalent source e.m.f.* in series with the input circuit. However, the power of the signal is equal to $E^2/4R$, where $E$ is the equivalent open-circuit voltage of the generator and $R$ the generator's internal resistance, both including the artificial aerial when used.

Thus, the value of the load must be specified in the tests. Moreover, when an artificial aerial is employed, the indicated output of the signal generator must be multiplied by an appropriate correction factor to obtain the equivalent source e.m.f. Unless care is taken over these points significant error of measurement can result.

## SENSITIVITY TESTS

The maximum sensitivity of a tuner (though more suitable for an f.m. radio or tuner–amplifier) is a measure of the r.f. input signal modulated 30% at 1kHz (BSI) required to produce a 'standard output' with volume and level controls at maximum. If one refers the 'standard output' to the completion of limiting, then a tuner providing the full-line curve in Fig. 3.2 would have a maximum sensitivity of 6μV.

In general, though, so far as receivers are concerned, the standard output is 50mW (BSI). Other values can be used provided they are specified. and the setup in Fig. 3.10, with S1/S2 at position B. is suitable for the sensitivity measurement.

### Noise-Limited Sensitivity

When the limiting is complete, or a standard output obtained, before the maximum S/N ratio is reached. the sensitivity can be referred to any earlier S/N ratio. This is called the noise-limited sensitivity. For example, if the output reaches limiting (or a standard value) from an input of 2μV, the S/N ratio at that input from the curve in Fig. 3.11 would be about 35dB.

Sometimes the input voltage is plotted as a function of the measuring frequency with a specified S/N ratio as the parameter, 20dB, 30dB or 40dB being ratios commonly used.

**Usable Sensitivity**

This is an IHF measurement, the test setup for which is given in Fig. 3.13. IHF standard f.m. test frequencies differ from BSI, being 90, 98 and 106MHz as a group or 98MHz when the test is performed at one frequency only. The standard test modulation also differs from 1kHz and 30%, being 400Hz and 100% equivalent to ±75kHz deviation.

The IHF artificial aerial for tuners with balanced input comprises a pair of resistors, one connected in series with each terminal of the signal generator, of such a value that the total resistance seen by the tuner, including the generator resistance, is 300Ω. However, the artificial aerials previously referred to (Fig. 3.12) would be suitable, and when used the generator output should be subjected to the correction factor already mentioned.

Going back to Fig. 3.13, the plan is to set SW1/SW2 to position B and to adjust the sensitivity of the audio millivoltmeter to produce a convenient readout datum when the tuner is accurately adjusted to a fairly strong r.f. signal carrying 400Hz modulation at 100%.

SW1/SW2 are then set to position A, which removes the 30dB attenuator and introduces a 400Hz 'null' filter. This 'notches out' the modulation signal and leaves only the hum, noise and distortion, and it is to this that the audio millivoltmeter now responds.

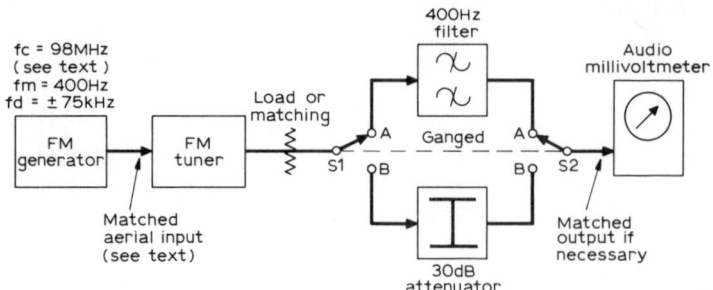

Fig. 3.13 Instrument setup for measuring usable sensitivity according to IHF. The filter is a 'null' type, corresponding to fd.

The internal filter of a THD test set is suitable for this filtering application.

The idea is then gradually to reduce the generator signal until the originally established reference datum is indicated by the audio millivoltmeter. The r.f. input signal required to secure this condition is the usable sensitivity. This is one of the most frequently adopted sensitivity tests for f.m. tuners.

## CAPTURE RATIO

This is another IHF test which reveals how well a tuner can reject an unwanted signal falling on the same frequency as the wanted signal, and it takes into account the functions of the f.m. detector, the limiter and automatic gain control circuits.

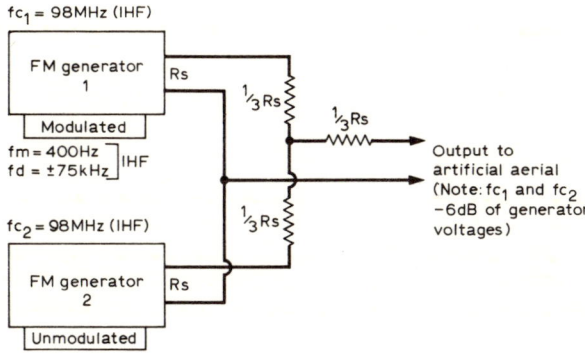

Fig. 3.14  Method of connecting two signal generators to provide a matched combined output. The common connection is 'earthy'.

The test requires two signal generators, but only one of them need embody facilities for f.m. The other can be an a.m. generator with the modulation switched off. The signals from the two are combined as shown in Fig. 3.14 (per BSI), and the combined output is taken to the tuner through a suitable artificial aerial (Fig. 3.12).

The two generators are tuned to the selected carrier frequency (98MHz IHF), and the test setup in Fig. 3.13 is used to set the modulated generator output to a value corresponding to 30dB usable sensitivity, based on the test procedure already described. While this is being done the output of the unmodulated generator is turned to zero.

### Test Set-up

The setup in Fig. 3.15 (or an equivalent) is adopted and, with the tuner controls set normally, the sensitivity of the audio millivoltmeter is adjusted to provide a convenient reference datum with SW1/SW2 set to position A and with the attenuator switched for 1dB attenuation.

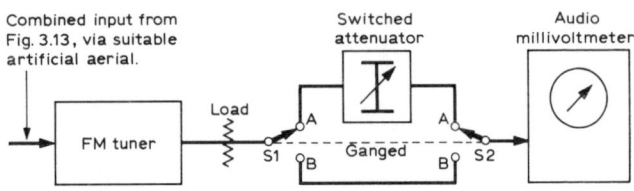

Fig. 3.15   Suggested instrument setup for measuring capture ratio (see text).

SW1/SW2 are then set to position B and the output of the unmodulated generator gradually turned up until the audio millivoltmeter again indicates the reference datum. The signal input to achieve this condition is noted.

SW1/SW2 are switched back to position A and the attenuator switched for 30dB attenuation. A reference datum is once more established on the audio millivoltmeter under this new condition by increasing its sensitivity.

SW1/SW2 are then switched to position B and the level of the unmodulated signal further increased until the audio millivoltmeter backs to the established reference datum. The signal input required to achieve this condition is also noted. The voltage ratio of the two noted signal values is converted to decibels, and the capture ratio is half of this dB value.

### Capture Effect

The *capture effect* is the ability of an f.m. tuner to suppress the weaker of two signals at the same frequency present at the input. The weaker one results in amplitude modulation of the stronger one, this then being removed by the limiter. There is no a.m. parallel of the effect.

In the IHF test just described the capture ratio refers to the ratio of the wanted to unwanted signal levels required for 30dB suppression of the unwanted signal when the two signals have exactly the same frequency and for 100% modulation.

## Tuner Tests

The test can be repeated for 30% modulation, or for any other depth, but that used should be specified. A curve is sometimes plotted of the capture ratio over a range of inputs—above the level required for 30dB usable sensitivity—at 20dB intervals.

The *rated capture ratio* is defined as the largest dB value obtained when tests are made at each of the IHF recommended carrier frequencies and when the output from the modulated generator is 1mV.

The tests are not particularly easy to handle, and care has to be taken to ensure that the two generators remain frequency coincident during the whole time that the tests are being performed. It should also be noted that the 'star network', used to combine the two signals, introduces 6dB attenuation.

## SELECTIVITY

There are various ways of measuring the overall selectivity of an f.m. tuner, which is defined as its ability to separate a wanted signal from an unwanted one on a near frequency.

Because selectivity is related to the nearness of the frequencies of two signals and on their strengths, the test should take these factors into account, and as a consequence several procedures require the use of two generators, as for the capture ratio test.

### Single Generator Method

However, a fair assessment of selectivity can be obtained by the use of one generator, using the setup in Fig. 3.16. The generator is matched to the aerial input of the tuner (by the use of an artificial aerial if necessary), the carrier set to one of the BSI recommended values (it is sometimes desirable to test at three or seven frequencies (Table 3.1)) and the modulation set to 30%. The tuner is then

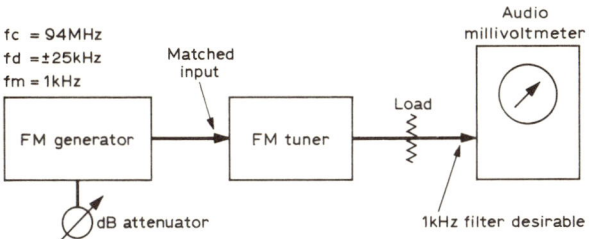

Fig. 3.16 Instrument setup for single generator selectivity test.

accurately adjusted to the signal and the audio millivoltmeter set to provide a convenient readout datum.

The idea is then to adjust the carrier frequency of the generator in convenient steps either side of the on-tune frequency and at each step to record the increase in r.f. signal level required to retain the readout datum.

It might be necessary to continue the measurements in 100kHz steps until the r.f. input signal ratio exceeds 80dB or until the input signal power exceeds 0dB(mW), corresponding to a generator e.m.f. of 1·1V into 300Ω and 550mV into 75Ω, whichever happens first. A plot of the selectivity can then be constructed as shown in Fig. 3.17.

Hum and noise sometimes make it difficult to obtain accurate measurements at the extremes of the response, and for this reason it is desirable to include a 1kHz bandpass filter in the readout circuit (see Fig. 3.18).

## Two Generator Method (BSI)

BSI and IHF both describe methods of selectivity measurement using two signal generators, the outputs of which are combined as shown in Fig. 3.13, but $f_{c1}, f_{c2}, f_m$ and $f_d$ differ from those indicated on this diagram of course. A suggested instrument setup for the BSI method is given in Fig. 3.18.

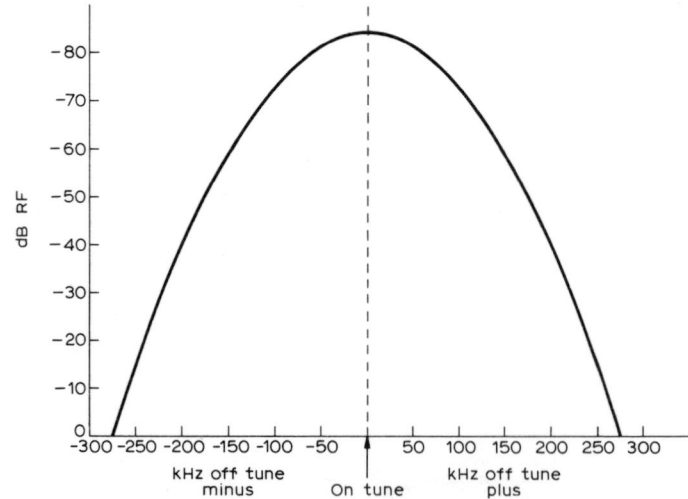

Fig. 3.17  Selectivity plot obtained with the setup in Fig. 3.16.

## Tuner Tests

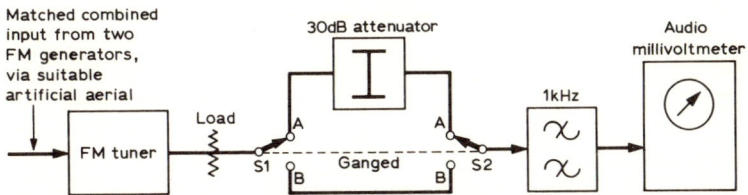

Fig. 3.18   Instrument setup for two generator selectivity test.

The generator simulating the wanted signal is modulated to 30% at 1kHz and its r.f. output is adjusted to that level required for the test. BSI give preferred values of input signal level, which are 1, 10 and 100µV and 1, 10, 100 and 1,000mV, with 1mV (i.e. 60dB(µV)) being the standard input signal level. These are all equivalent open-circuit voltages.

The wanted signal is tuned in, and in the case of f.m. receivers and tuner-amplifiers the output power is adjusted to avoid a.f. overload. During this time the unwanted signal is turned to zero, SW1/SW2 set to position A, and the audio millivoltmeter adjusted to provide a convenient readout datum.

The modulation of the wanted signal is then turned off, and the level of the unwanted signal, modulated to 30% at 1kHz, is gradually increased, with SW1/SW2 set to position B, until the audio millivoltmeter reads the previously established reference datum.

The dB ratios of the wanted-to-unwanted signal levels are plotted as functions of points corresponding to frequency differences of at least ±100, 200, 300, 400 and 600kHz between the wanted and unwanted signals, with the wanted signal as the parameter, as shown in Fig. 3.19.

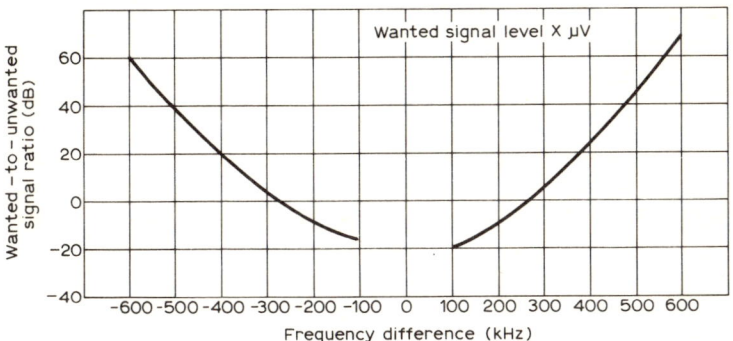

Fig. 3.19   Selectivity curves using the two signal generator method (see text).

Curves are constructed at the various BSI test frequencies (Table 3.1) and at the preferred levels, as mentioned. This is not a particularly easy test and errors can result from insufficiently sophisticated test equipment. The generators should be accurately calibrated and that simulating the unwanted signal should be easily tunable over intervals of 100kHz. The 1kHz bandpass filter shown in Fig. 3.18 is desirable to avoid readout errors due to hum and noise.

### Two Generator Method (IHF)

This test is based on the setup used for the capture ratio test (Fig. 3.15). In this case, however, the generator producing the unwanted signal is separated in frequency from that producing the wanted signal by 200kHz, corresponding to a standard f.m. channel. The wanted signal is unmodulated and is fed to the tuner at the level required for 30dB usable sensitivity.

The generator simulating the unwanted signal is modulated to a depth of 100% at 400Hz (note that this differs from BSI) and its r.f. output is gradually increased from zero until the readout is 30dB below that established initially as a reference datum (see under *Capture Ratio*).

The ratio of the signal levels from the two generators is expressed in decibels which is the value for 100% modulation *adjacent channel* selectivity. The tests can be made at 30% modulation, giving 30% modulation adjacent channel selectivity, and they can be repeated with the separation at 400kHz to provide a 100% or 30% readout in terms of *alternate channel* selectivity.

Obviously, the selectivity ratios will differ significantly between adjacent and alternate channel selectivity and 100% and 30% modulation, meaning that tuner specifications not providing a clear indication of this parameter are of little comparative value.

## SPURIOUS RESPONSES

### Amplitude Modulation Suppression

One test to measure the ability of an f.m. tuner to suppress amplitude modulation requires a single generator whose modulation can be switched over amplitude and frequency.

Based on the BSI procedure, the generator is first set up to provide an f.m. signal, 30% at 1kHz, at the selected carrier frequency and the tuner is accurately adjusted to this. The output load is connected via a switched (1dB step) attenuator to an audio millivoltmeter, and a 0dB datum is established on the meter by adjusting its sensitivity when the attenuator is set to about −40dB.

The modulation is then changed to amplitude at 30% and 1kHz, and the attenuator switched down until the datum reading is re-established. The amount of decibels switched down is a measure of the a.m. suppression ratio at the r.f. input and carrier frequency selected.

**Input Levels**

BSI recommend that tests be made at input levels between 31·6µV and 10mV (30dB(µV) and 80dB(µV) respectively) and at points up to ±100kHz relative to the tuned frequency (over a greater range than this if automatic frequency control—a.f.c.—is active). In a well designed tuner maximum suppression occurs when the tuning is accurate, but the suppression should not diminish too rapidly over small tuning errors, since this would imply that the limiting falls off too quickly.

This sort of test fails to appraise the suppression under conditions of transient-type interference, so a tuner may have good suppression relative to steady-state amplitude modulation and mediocre suppression to transient-type amplitude modulation.

One major problem with this test lies in residual frequency modulation on the amplitude modulated carrier. Inexpensive generators are totally unsuitable since they are likely to yield a high deviation of f.m. even when switched to a.m.—so beware!

The IHF test for this parameter is similar, but 100% f.m. at 400Hz is adopted, and tests are made at various points over the modulation frequency range of 30–15,000Hz.

**Image Rejection Ratio**

There are numerous tests designed to express the immunity of an f.m. tuner to spurious input signals at frequencies removed from the tuned frequency.

A particularly important one is the image response, resulting from the responsiveness of a superhetrodyne tuner to two frequencies whose difference from the local oscillator frequency is equal to the intermediate frequency. Only one of these is the wanted one. The other is the image response, sometimes called the second channel response.

Based on BSI, the rejection ratio is obtained by first measuring the tuner's maximum sensitivity (see under *Maximum Sensitivity*, page 85) and then by measuring the effective sensitivity at the image frequency, securing the same readout in both cases. The image rejection ratio is then equal to the ratio in decibels of the levels of the unwanted (image) to wanted signals.

To obtain the accurate image frequency it is necessary to feed to the tuner a very strong signal initially at a frequency approximately corresponding to that of the image response, and then to tune the generator until the exact tuning point is established. Tests are usually made at three or more frequencies, based on the BSI recommendations (Table 3.1, page 74).

**Image Interference Ratio**

Measurement of the image *interference* ratio is different. For this test two signal generators are used, one for the wanted signal and the other for the unwanted (image) signal. That tuned to the wanted signal is modulated to 30% at 1kHz, and the tuner is adjusted to this to secure a readout reference datum, while the other generator is turned to zero r.f.

The modulation on the wanted signal is then turned off, and the other generator, modulated to 30% at 1kHz, is increased in r.f. output until the readout is 30dB below the previously established datum. The setup in Fig. 3.18 can be used for this test. The image interference ratio is equal to the ratio in decibels of the levels of the unwanted (image) to wanted signals.

**Intermediate Frequency Rejection Ratio**

Superhetrodyne tuners are also responsive to signals falling at the intermediate frequency, and the ratio is measured in the same way as described for the image, except that the input is adjusted to the intermediate frequency.

The intermediate frequency *interference* ratio is also measured in the same way as the image *interference* ratio, except that the generator producing the unwanted signal is tuned to the intermediate frequency.

## OTHER SPURIOUS RESPONSES

F.M. tuners can exhibit other spurious responses, which can be located by slowly adjusting the generator over Band II when it is delivering a high signal level and noting the frequencies at which outputs occur. Generators which rely on harmonics of lower

## Tuner Tests

frequencies for their r.f. output signals are useless for this purpose.

The measurement of a rejection ratio (as distinct from an interference ratio) is merely based on discovering the ratio of the sensitivity through the main channel of the tuner at a given r.f. input level, modulation frequency and deviation (i.e. modulation percentage) to that of the sensitivity through the spurious channel under similar conditions, the ratio (in terms of signal voltages) then being converted to decibels.

**Intermodulation Components**

Some apparently spurious responses arise from intermodulation components developing in the r.f. and frequency changer or mixer stages of f.m. tuners under conditions of strong input signals. Tuners with poor pre-mixer selectivity are more prone to this trouble than counterparts in which two or more variably tuned circuits precede the mixer.

Since the intermodulation components result from circuit and active device non-linearity, the more linear the circuit or active device, the fewer the spurious responses from this cause.

**Third Order IMD**

One particularly troublesome interference potential is third-order intermodulation due to three programme signals passing simultaneously through the non-linear front-end yielding signal $f_s$ from signals $f_2 + f_4 - f_3$. where these correspond respectively to the frequencies of *Radios 2, 4* and *3* of a station group. In this order, $f_s$ lies in the channel occupied by *Radio 3*.

Thus if intermodulation is occurring in the tuner the wanted *Radio 3* will be accompanied by the spurious signal $f_s$, and since this is affected by the modulation of any of the three signals, the interference on tuned *Radio 3* manifests as a kind of 'warbling' sound.

This happens more often when the front-end transistors are pushed well towards non-linearity due to very strong aerial signals. Some of the more recent f.m. tuners equipped with field effect transistors and two or more variably tuned circuits in front of the mixer stage are less troubled by this.

This is because FETs can handle a greater signal level than bipolar transistors before running hard into non-linearity and also—most important—because the transfer characteristic of an FET is almost exactly square law, meaning that it produces significantly less third harmonic than a bipolar transistor whose characteristic contains third-order terms.

## Non-linearity

Non-linearity can also produce a spurious signal $f_s$ due to signals $f_1$ and $f_2$, such as $f_s = f_1 + f_2$. Signal $f_s$ can then beat with a harmonic of the local oscillator to yield the intermediate frequency. Thus when a tuner so troubled is adjusted so that the oscillator harmonic with $f_s$ gives the i.f., the response will carry both signals $f_1$ and $f_2$ which, if modulated, will be heard together.

Other combinations of signal and oscillator harmonics can also produce similar interference, generally called 'double beat interference'. Another is 'repeat spot interference' which results from an intermodulation component produced at the front-end beating with a harmonic of the local oscillator to produce the i.f.

## Beat Interference

Still another is 'continuous beat interference' which arises from an intermodulation component produced by two strong signals falling within the i.f. passband. The intermodulation component might be related to the fundamental or to the harmonics of the two signals causing it. Interference of this kind is not affected by the tuning.

While some of these effects are not directly caused by a real spurious responses, they should be understood by the audio technician because they can result when two signals are applied to a tuner under test and when high levels are adopted. Moreover, it may well be necessary for a technician to simulate the conditions responsible for intermodulation interference, for example, as a means of locating it and (hopefully) alleviating it.

# AUTOMATIC FREQUENCY CONTROL

Tuners equipped with automatic frequency correction (or control) should be tested with the control both on and off, and this applies particularly with regards to such parameters as usable sensitivity, capture ratio, frequency response and distortion.

The control itself can be tested but as this requires rather specialised signal generators few operators are able to secure truly meaningful results.

## Pull-in Range

The pull-in range of the control can be measured in the following way. Set the tuner to a recommended (BSI) carrier frequency

(Table 3.1), and adjust a signal generator connected to the tuner and delivering an unmodulated carrier at a level of 1mV as accurately as possible to the actual frequency to which the tuner is adjusted.

Next, take a second generator tuned to the tuner i.f. and couple the output of this very loosely to the tuner, then very carefully adjust the generator tuning for dead beat (assuming the tuner will thus respond). To detect this, of course, it is necessary for the tuner to be working into an amplifier and loudspeaker system.

The next move is to detune the generator delivering signal to the tuner by a small amount and then note the frequency difference required to secure the dead beat condition again on the loosely-coupled generator. The degree of retuning needed corresponds to the a.f.c. error.

The process should be continued first on one side of the correct tuning point and then on the other side until the limit of the control range is reached, which is expressed in $\pm$ kHz with reference to the level of the input signal and the test frequency used.

The IHF method is similar except that the pull-in range is referred to the difference between the frequency of the generator and the tuned frequency of the tuner required to reduce the tuning error to 22·5kHz when the input signal is at the level corresponding to 30dB usable sensitivity.

## F.M. STEREO TESTS

For serious work on stereo f.m. tuners an f.m. stereo generator is required. Such an instrument has an r.f. section similar to that of an ordinary f.m. generator. Circuits are included, however, to modulate left and right stereo channels, based on the 19kHz pilot tone system, on to the single carrier. It may be possible to apply modulation tone of one frequency to the left channel and of a different frequency to the right channel.

### Decoder Performance

With both channels so modulated it becomes a simple matter to appraise the performance of a stereo decoder under two-channel dynamic conditions. Moreover, with one channel only modulated, crosstalk in the non-speaking channel can easily be measured. The idea is to establish a suitable output datum on the speaking channel and then to measure the level of signal that this produces in the non-speaking channel in terms of a dB ratio.

Thus, if we measure, say, 1mV at the output of the speaking

channel and 0·1mV at the output of the non-speaking channel the separation would be 20dB. The stereo tuner specification almost always gives this at 1kHz modulation frequency at least.

**Separation Test**

When the multiplex generator has facilities for switching (or tuning) to different modulation frequencies, then the separation test can be performed at low and high frequencies, the ratio generally diminishing at the latter due to capacitive effects and the limitations inherent to stereo decoders.

Most stereo tuners have a crosstalk performance of, at least, 20dB at 1kHz. Top quality models can go as high as 36dB, falling possibly to about 26dB at 10kHz.

The separation can be optimised either by adjusting the crosstalk preset relative to the left and right channels, if such is fitted, or by carefully adjusting the pilot tone (and sometimes the reference frequency) tuned circuit while the crosstalk signal in the non-speaking channel is being measured.

Audio technicians who are not adverse to late-night working can use the BBC's stereo tests to appraise and adjust stereo tuners. These occur after close-down from *Radio 3* transmitters equipped for stereo broadcasting, and generally continue to seven minutes to midnight.

The test schedule at the time of writing and as provided by the BBC is given in Table 3.2.

**Tone Tests**

The special tone transmissions on Wednesdays and Saturdays facilitate certain checks on stereo tuners. For example, test 1 provides identification of the left and right channels, test 2 allows adjustment to the phase of the 38kHz subcarrier signal, test 3 provides a check of distortion with the signal wholly in the L–R stereo channel, test 4 is similar but with the signal wholly in the L + R mono channel.

Test 5 provides a check of distortion with the signal equally divided between the stereo and mono channels, test 6 is suitable to check left to right crosstalk, test 7 to check right to left crosstalk, test 8 allows a check of left channel frequency response and left to right crosstalk at high frequencies.

Test 9 allows a check of right channel frequency response and right to left crosstalk at high frequencies, test 10 provides a reference level for the noise test which is provided by test 11.

## Tuner Tests

### Table 3.2
### STEREOPHONIC BROADCASTING TEST TONE TRANSMISSIONS

*Every day except Wednesday and Saturday*

To facilitate channel identification and adjustment of channel cross-talk, 250 Hz tone is transmitted in the left channel only from about four minutes after the end of Radio 3 until 2355. This test may be interrupted from time to time.

*Wednesday and Saturday*

| Test | Time | Left Channel (A) | Right Channel (B) |
|---|---|---|---|
| 1 | 23.30 | 250Hz at zero level | 440Hz at zero level |
| 2 | 23.32 | 440Hz at zero level | 440Hz at zero level, antiphase to left channel |
| 3 | 23.35 | 440Hz at +8dB | 440Hz at +8dB, antiphase to left channel |
| 4 | 23.37 | 440Hz at +8dB | 440Hz at +8dB, in phase with left channel |
| 5 | 23.39 | 250Hz at +8dB | 440Hz at +8dB |
| 6 | 23.40 | 250Hz at zero level | Nothing |
| 7 | 23.44 | Nothing | 440Hz at zero level |
| 8 | 23.47.20 approx. | Tone sequence at −4dB: 60Hz, 900Hz, 5kHz, 10kHz. This sequence is repeated. | Nothing |
| 9 | 23.48.20 approx. | Nothing | Tone sequences as for left channel at 23.47.20 |
| 10 | 23.49.20 | 250Hz at zero level | Nothing |
| 11 | 23.51 | Nothing | Nothing |
| | 23.53 | Reversion to monophonic transmission | |

**Notes**
1. This schedule is subject to variation to accord with programme requirements and essential transmission tests.
2. The zero level reference corresponds to 40% of the maximum level of modulation applied to either stereophonic channel before pre-emphasis. All tests are transmitted with pre-emphasis.
3. Periods of tone lasting several minutes are interrupted momentarily at one-minute intervals.

## Crosstalk Presets

In a tuner with separate presets for the adjustment of sub-carrier phase (i.e. core of tunes circuit) and crosstalk (usually resistive preset), the subcarrier phase should first be adjusted using test

2 to produce maximum output from either the left or right channel, after which the crosstalk (may be called 'separation') preset should be adjusted for the least crosstalk, using tests 6 and 7. Tuners carrying no crosstalk or separation preset should be adjusted for the least crosstalk by the pilot tone or subcarrier phase tuning core, using tests 6 and 7.

**Instruments**

While one cannot deny the usefulness of these tests, not all technicians are keen on tackling stereo decoder adjustments at midnight (it may be different for the enthusiastic amateur), so an f.m. stereo generator is really as essential for the audio technician as a colour generator is for the television service technician.

A number of such instruments is available, but, as would be expected, quite costly. A relatively inexpensive instrument which may be of interest, however, is the IG-37 Heathkit. This is from the Heath Company at Benton Harbour, Michigan, but is available from Heath (Gloucester) Limited, and in kit form costs £40·50. It is pictured in Fig. 3.20.

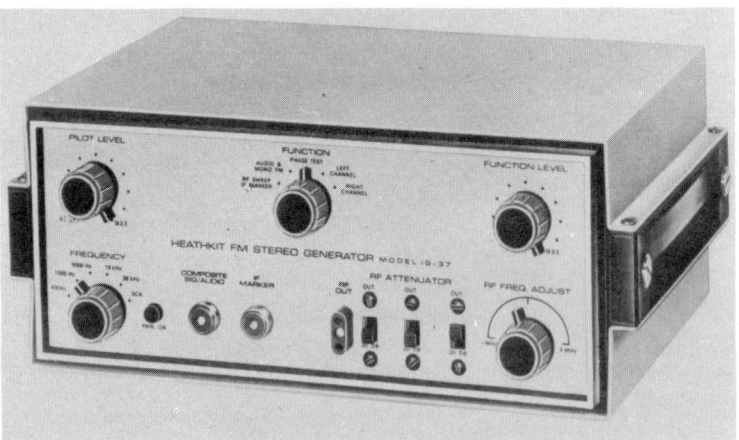

Fig. 3.20 Heathkit f.m. stereo generator, Model IG-37. This is described in the text.

The carrier is nominally 100Hz, adjustable approximately ±2MHz, while the 19kHz pilot tone is ±2Hz, under crystal control. It produces mono f.m. or composite stereo f.m. signals via an r.f. attenuator, switchable in 20dB steps to 60dB. The audio is switchable over 400Hz, 1kHz and 5kHz, and the level of the pilot tone is

# Tuner Tests 101

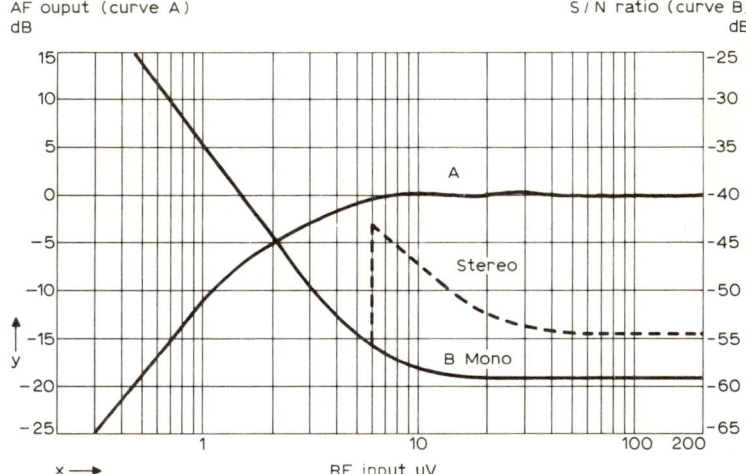

Fig. 3.21 Curves obtained from a stereo tuner. A is limiting and full-line B the mono S/N ratio. The stereo S/N ratio is shown by the broken-line curve.

adjustable from zero to the 10% maximum, which is useful for checking switching levels and lock-in range, etc.

In addition to providing the signals required for most tests on stereo tuners, the instrument also incorporates a 100MHz sweep signal on a clear part of Band II for r.f. and i.f. alignment and a crystal-controlled marker generator suitable for i.f. and dial tracking tests.

## FM STEREO S/N RATIO

Owing to the action of the stereo multiplex and the wider bandwidth required for this signal, compared with mono, the S/N performance of a stereo tuner rarely approaches that of the same tuner running mono.

To check the stereo S/N ratio, therefore, it is necessary to employ a stereo encoded generator. The measuring procedure is similar to that already described for mono tuners (page 82). Some idea of how the stereo S/N ratio falls short on the mono S/N ratio can be gleaned from Fig. 3.21.

Curve A shows the limiting performance, which is the same on both mono and stereo, while curve B shows the mono S/N ratio, compared with the stereo S/N ratio shown by the broken-line curve. It will be seen that while the mono S/N ratio is about 55dB (referred to 100% modulation) at an r.f. input of about 6μV, the stereo S/N ratio at that input is less than 45dB.

**Switching Point**

The sudden rise of the stereo curve indicates the switching point of the stereo decoder, the switching happening in this case at 6μV. Most stereo tuners nowadays incorporate automatic switching at a specific level (sometimes preset), this being activated by the presence of pilot tone. Note also that while the ultimate mono S/N ratio is almost 60dB, the ultimate stereo S/N ratio is only about 55dB.

This 5dB difference is the cost of the wider band encoding, etc. The difference is less with some tuners, but it is always greater at low r.f. signal input levels (about 12dB at 6μV in the diagram), which reveals the need for the highest aerial signal input for the least noise on stereo.

Sometimes the f.m. muting is geared to the stereo switching, in which case the audio remains muted until the input signal is of a sufficient level to switch the stereo decoder. More practical information on these points is given in my companion book, *Tuners and Amplifiers*, by the same publisher.


## A.M. SECTIONS

Since some quality tuners and tuner-amplifiers are equipped with facilities for the reception of a.m. signals in one or more of the appropriate bands, the audio technician will from time to time find a need to conduct tests in this area, which is fringing on that possibly more in the province of the radio service technician.

Nevertheless, more detailed testing than that required for servicing might be called for to secure an absolute appraisal of some of the a.m. parameters. In general, some of the testing procedures already described for f.m. will be applicable also for a.m., based on different frequencies and modulation, of course.

**Test Frequencies**

BSI recommend three standard groups of measuring frequencies as detailed in Table 3.3 (right), while IHF recommend frequencies of 600, 1,000 and 1,400kHz, taking 1,000kHz as the 'standard' when measurements are to be made at one frequency only.

## Tuner Tests

Table 3.3

| Eighteen | Eight | Three |
|---|---|---|
| 160kHz | — | — |
| 210kHz | 210kHz | 210kHz |
| 250kHz | — | — |
| 550kHz | — | — |
| 600kHz | 600kHz | — |
| 800kHz | — | — |
| 1MHz | 1MHz | 1MHz |
| 1·2MHz | — | — |
| 1·4MHz | 1·4MHz | — |
| 1·6MHz | — | — |
| 6·1MHz | — | — |
| 7·2MHz | 7·2MHz | — |
| 9·6MHz | — | — |
| 11·8MHz | 11·8MHz | 11·8MHz |
| 15·3MHz | — | — |
| 17·8MHz | 17·8MHz | — |
| 21·6MHz | — | — |
| 25·8MHz | 25·8MHz | — |

### Test Signal Voltages

The preferred input voltages for open aerial tuners (receivers) per BSI are from 0dB(μV) to 120dB(μV) in 20dB steps, corresponding to equivalent open-circuit voltages from 1μV to 1V, with intermediate values spaced at 10dB(μV) intervals, including 130dB(μV), corresponding to an equivalent open-circuit voltage of 3·16.

Preferred inputs for tuners (receivers) with ferrite rod aerials are 0dB(μV/m) to 120dB(μV/m) in 20dB steps, corresponding to equivalent fields from 1μV/m to 1V/m, with intermediate values spaced at 10dB(μV/m). 10mV and 10mV/m are standard inputs.

### Artificial Aerials

The BSI a.m. artificial aerial is given in Fig. 3.22, (a) for a single generator input and (b) when the signals from two generators need to be combined for the test.

### Coupling to Ferrite Rod Aerials

For coupling signal to a ferrite rod aerial to obtain meaningful measurements, an electrostatically-screened coaxial test coil is required, the suggested design of which is shown in Fig. 3.23.

Orientation should be such that the axis of the ferrite rod is normal to the plane of the test coil when its plane falls at the centre of the rod, as shown in Fig. 3.24.

The equivalent signal field intensity is then $188 \cdot 5 N A^2 I / 2 Y^3$, where $N$ is the number of turns of the coil, $A$ is the radius of coil in metres, $Y$ is the distance in metres between the coil axis and the ferrite rod (see Fig. 3.24) and $I$ is the coil current.

When it is necessary to induce two signals at different frequencies into a ferrite rod, either two test coils can be employed, one set up each side of the ferrite rod, or the two signals can be applied to one coil, as shown in Fig. 3.25.

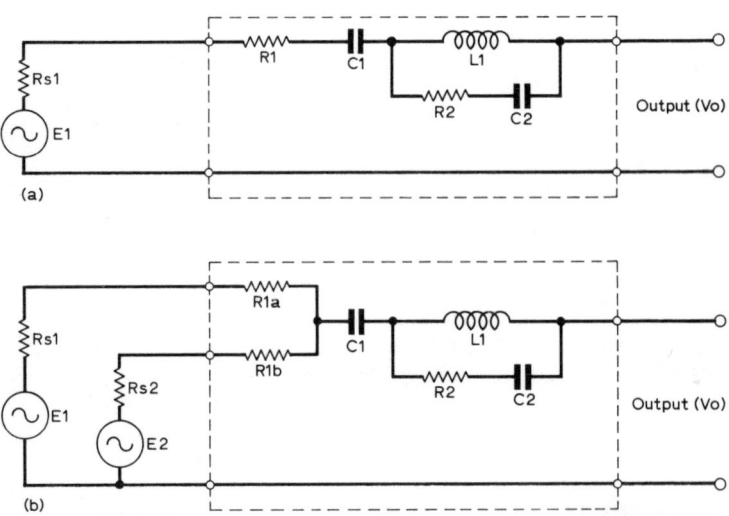

Fig. 3.22  A.M. artificial aerials. (a) for single generator connection and (b) when the test requires the combined signals from two generators. In both cases, C1 is 125pF, C2 400pF, L1 20μH and R2 320 ohms. At (a) $R_{s1}$ is the source resistance of the generator, E1 the generator e.m.f. and R1 such that $R_{s1}$ + R1 is 80 ohms. At (b) $V_o$ is E1 + E2/2, $R_{s1}$ and $R_{s2}$ source resistances of the two generators, R1a = R1b such that $R_{s2}$ + R1b is 160 ohms.

In Figs. 3.24 and 3.23, R is selected to adjust the signal current so that a signal field corresponding to one-twentieth of the generator e.m.f. is produced when $Y$ is 0·6m. The value can be determined from the field strength expression by taking the generator output to be 1V and solving for $I$ when the values pertaining to the physical parameters of the coil (Fig. 3.23) are substituted and $Y$ is taken as 0·6m. For different values of $Y$, of course, similar calculations will

# Tuner Tests

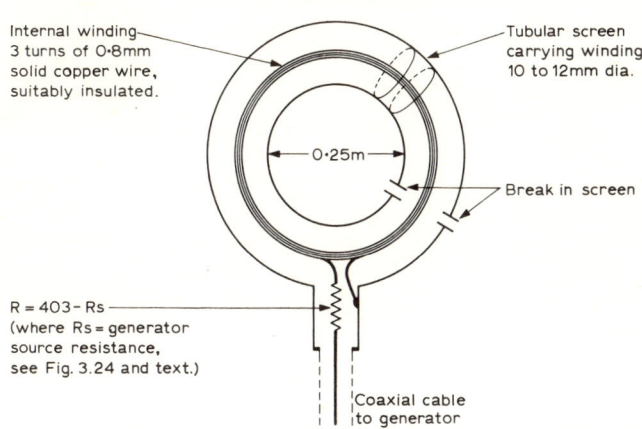

Fig. 3.23 Coil for injecting signal to a ferrite rod aerial.

be necessary. It is possible to calculate *I* for the expression when the the coil impedance is known, based on the voltage across the coil.

When the two generators are employed as in Fig. 3.24, R1 and R2 have the same value when $R_{s1} = R_{s2}$, which is usual, and the same method of calculation applies. However, in Fig. 3.25, E is equal to E1 + E2.

Since the signal field is geared to the coil current it is obviously necessary to take into account the source resistance, which is why this is subtracted from the series resistance to give the actual value of the resistor required. Shunt capacitances limit the top frequency when one generator is used to about 3MHz and to about 1·6MHz when two generators are used. The lowest frequency is 150kHz.

The IHF artificial (or dummy) aerial consists merely of a 200pF capacitor, approximating an open, single wire aerial (including the lead-in) of 4m effective height over the frequency range 540 to 1,600kHz.

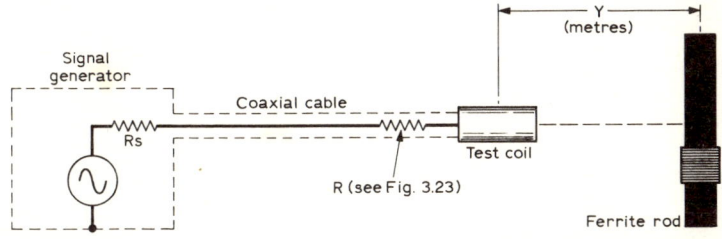

Fig. 3.24 Showing the position of the ferrite rod aerial relative to the test coil.

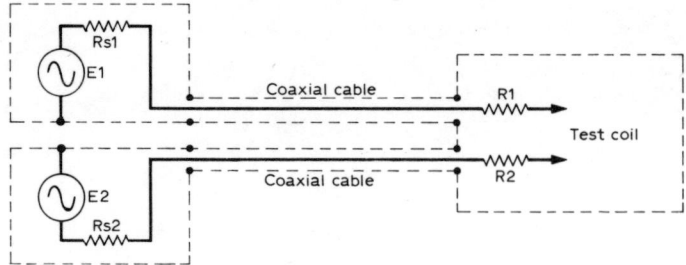

Fig. 3.25 Showing how two signal generators can be coupled to a common test coil (see text).

## Test Modulation

Both BSI and IHF indicate a modulation depth of 30% for a.m. tests, with 1kHz modulation frequency BSI and 400Hz IHF. It is not intended to run through all the a.m. tests which, as I have already intimated, are often similar to those of the f.m. parameters but at different frequencies and modulation.

## Usable Sensitivity

An example is the IHF usable sensitivity. Here the test is similar to f.m. except that the input signal is reduced to the least value that produces a 20dB (30dB f.m.) rise in the indicated output compared with the output readout obtained through a 400Hz 'null' filter. See Fig. 3.13 for the f.m. instrument setup.

## S/N Ratio

The BSI procedure for measuring the a.m. signal-to-noise ratio is similar to that described on page 82 for measuring the f.m. S/N ratio. The instrument setup is also similar to that shown in Fig. 3.10, except for the modulation and carrier frequency.

## Volume Sensitivity

This IHF test refers (as a dB ratio) the tuner output to 1V when the modulated r.f. input is 100mV. The generator output is then reduced to the value which diminishes the audio output by 20dB. The results are expressed in µV of r.f. signal. When the tuner is equipped with sensitivity and selectivity controls, the tests are usually made at various settings of these.

**Frequency Response**

The IHF procedure stipulates an r.f. input of 5mV, the generator being modulated from an audio source variable from 30Hz to 15kHz. The response is that between the $-3$dB points indicated by an audio readout in the usual manner. If a selectivity control is fitted this is set to the least selectivity, and the bandwidth of any filter used is also stated.

These, then, are some a.m. test examples. It will be obvious to the reader how those described for f.m. can be adapted for a.m.

## TUNER–AMPLIFIERS

The tuner section of a tuner–amplifier can either be tested as a tuner in essential isolation, taking the audio output from the sockets (left and right in the case of stereo) delivering signal for tape recording, or as a complete receiver, in which case the output is taken from the main amplifier, across appropriate loads as required for amplifier tests (Chapter 2).

There are some differences in interpretation of the test results, depending on the scheme adopted, and these are examined in Chapter 5.

CHAPTER FOUR
# DISC PLAYING EQUIPMENT TESTS

DISC PLAYING EQUIPMENT consists of three primary components: (1) the pickup cartridge, (2) the arm and (3) the turntable unit, and it is often necessary for the audio technician to test these both separately and collectively.

## PICKUP TESTS

The three main parameters of a pickup cartridge are (i) trackability, (ii) frequency response and (iii) stereo separation. Because (ii) and (iii) can be affected by the playing weight, and because the playing weight is a function of the tracking ability, (i) is best tested first.

**Trackability**

The reactive forces developed between a recorded groove and the stylus working therein need to be combated by a downward force, which is often called the playing weight (or force). The greater the reactive forces, the greater the playing weight required to keep the stylus tip in faithful communication with the groove modulation. When the tip fails accurately to follow the modulation in all of its detail, the cartridge is regarded as mistracking.

The reactive forces are a function of the velocity and hence frequency and amplitude of the groove modulation and of mechanical factors of the cartridge, which are mechanical resistance (used for damping), compliance (the reciprocal of stiffness and necessary for the tip to follow large amplitude excusions) and effective tip mass.

The tracking ability of a cartridge can be related to the mechanical *impedance* at the stylus tip because this is composed of all the factors just named, just as electrical impedance is composed of resistance, capacitance and inductance, which respectively are analogous to mechanical resistance, compliance and mass.

It can be shown that mechanical impedance is also equal to the

*Disc Playing Equipment Tests* 109

force threshold divided by the velocity of the groove modulation (there is much more about this in my companion book *Pickups and Loudspeakers*, by the same publisher). Thus, if we can devise a scheme which will detect when the stylus fails to follow the modulation, we can gradually reduce the playing weight and calculate the mechanical impedance.

**Equipment Required**

This, in fact, represents a common test method for trackability. Essential items are a top-flight arm with minimal friction at the bearings (preferably not greater than a force at the stylus tip corresponding to 30mg in both the lateral and vertical planes. See under Testing Arm Bearing Friction, page 125) and with a 'flat' on the headshell to facilitate the positioning of laboratory weights, a set of laboratory weights for applying the tracking weight at intervals of 50mg once the 'average' tracking weight has been established.

The turntable unit must be of high quality and fixed, with the arm, to a solid motor board. The motor and drive system should introduce the least rumble and be of constant and accurate speed (i.e. minimal wow and flutter), preferably with a speed control and means of setting the speed accurately.

The cartridge for test should be carefully aligned and adjusted in the headshell for the least vertical tracking error and the arm (cartridge/headshell) adjusted for the least lateral tracking error (see the *Pickups and Loudspeakers* book) and the motor board should be levelled.

**Test Set-up**

The plan is then to feed the cartridge signal through an amplifier and filter system to an oscilloscope and audio millivoltmeter. Most audio millivoltmeters have an output socket, so the Y input of the oscilloscope can be connected to this. The instrument setup is given in Fig. 4.1.

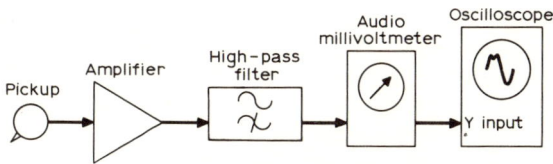

Fig. 4.1   Instrument setup for tracking tests.

The high-pass filter is included to remove the effects of rumble and disc ripple which, if present, will make it very difficult to examine the displayed waveform and read the audio millivoltmeter. Its −3dB point should occur around 100Hz and its ultimate rate of roll-off approach 18dB/octave.

It is also sometimes desirable to employ a low-pass filter to remove high-frequency noise, depending on the frequency of the test and on the high-frequency performance of the cartridge; but it should not be allowed to interfere with the readout up to about 10kHz.

In most cases the cartridge itself will act as a low-pass filter, and this effect can, in fact, inhibit the display of mistracking at high frequencies. Generally, however, it is possible to detect the mistracking threshold at frequencies up to 5kHz, but above this both patience and experience are vital ingredients!

**Test Records**

Recorded bands of single frequency over the range 300Hz–10kHz are required, and the recorded velocity of each band must be known. CBS test records are available with constant velocity bands from 500Hz to 20kHz and constant amplitude bands from 40Hz to 500Hz. The STR-5002-C and STR-100 discs are recorded like this (constant tone bands) in certain sections.

When only the amplitude or velocity of a recording is known, the unknown quantity can easily be found since the two are related such that the maximum transverse velocity is equal to $2\pi f A$, where $f$ is the frequency in Hz and A the peak amplitude. The velocity is generally quoted as a r.m.s. value, and is given by $1.41\pi f A$. Care must be taken to avoid intermixing r.m.s. and peak values in any such calculation.

The amplitude is given in cm and the velocity in cm/S. Thus, if the recorded amplitude is, say, 0·002cm and the frequency 1,000Hz then, using the above (peak amplitude) expression, the velocity is close to 12·56cm/S. Conversely, if the velocity (the known quantity) is, say, 10cm/S and the frequency 500Hz we can find the amplitude by altering round the expression to give $A = V/2\pi f$, which works out to almost 0·0032cm.

The recorded bands should not be of too high a velocity otherwise some of the less exacting cartridges will fail to track them anyway! There might also be 'curvature overload', which results when the stylus tip radius is too great to follow the sharp curves of the modulation at high frequencies and velocities.

The CBS disc just mentioned is recorded in the constant velocity bands to 0dB which, in this case, refers to 3·54cm/S r.m.s. However,

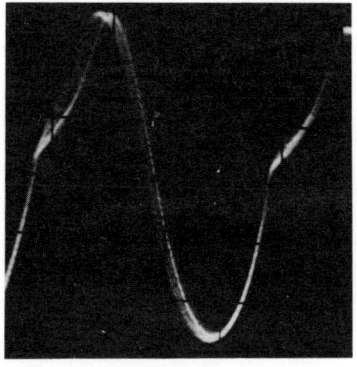 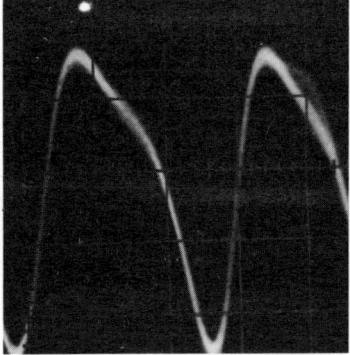

Fig. 4.2  Mistracking waveforms (a)—left—at low frequencies and (b)—right—at middle-high frequencies.

bands recorded up to 5cm/S should be suitable over the range 500Hz to 10kHz, with bands of smaller velocity outside these limits.

### Minimum Tracking Weight

So much, then, for the test discs; but what about the tests themselves? Well, these merely boil down to finding the least tracking weight required by the cartridge at each test frequency to maintain accurate tracking as indicated by the oscilloscope display and audio millivoltmeter readout. Useful test frequencies are 300Hz, 1kHz, 5kHz and (if a meaningful readout can be obtained) 10kHz.

Mistracking at low and middle frequencies is obvious by the mutilation of the oscilloscope display, as shown by the oscillograms in Fig. 4.2, where (a) is low-frequency mistracking and (b) middle-high frequency mistracking. The idea is to increase or decrease the playing weight as accurately as possible by the laboratory weights (placed on top of the 'flat' of the headshell carrying the cartridge over the stylus) until the waveform display loses its obvious mistracking symptom, and then to note the frequency and weight for this condition.

At frequencies above about 5kHz the low-pass characteristic of the pickup makes it less easy to define the mistracking threshold by the oscilloscope display alone. The audio millivoltmeter is then used in conjunction with the oscilloscope, since this will often show high-frequency mistracking by wavering of its pointer at the threshold.

The idea is to add just a little more tracking weight until the readout stabilises. It is often found that the output falls slightly when the cartridge is on the verge of mistracking, so it is necessary to add more weight to get a rise. With care, it is possible to do this while the pickup is playing the test band.

**Stylus Tip Impedance**

We now have three known factors: (i) velocity, (ii) frequency and (iii) tracking weight, so we can easily calculate the stylus tip impedance at each test frequency by dividing the tracking weight by the velocity. However. since we are dealing in *force* rather than weight the tracking weight has to be converted to the unit of force, which is the dyne, mechanical impedance in 'ohms' then being dyne/cm/S. The dyne as unit force corresponds to the pull of gravity on 1/980 gram at latitude 45 degrees and at sea level, but for most practical purposes we can take the dynes as 1/1000 gram or 1mg.

Thus, a test cartridge will have a mechanical impedance of 100Ω when the tracking threshold of 5cm/S modulation occurs at a force of 500 dynes (equal to about 0·5g). The impedance will not be the same for all frequencies, of course, which is why it is necessary to check at several frequencies over the spectrum; but the better the tracking ability of a cartridge, the lower will be the mechanical impedance at its tip.

It is possible to extrapolate and find approximately the velocity at the test frequency that a cartridge will track at its recommended tracking weight (or any other weight, come to that, provided this does not exceed the maximum specified by the manufacturer, since then the 'mechanics' would be likely to bottom). For example,

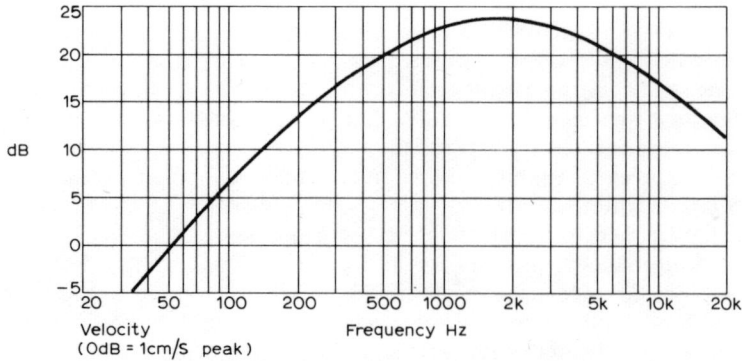

Fig. 4.3 Estimate of peak levels recorded on a music disc. However, it is likely that some discs have transient peaks rising above the levels indicated.

assuming perfect mechanical linearity a cartridge with a tip impedance of 100Ω calculated at 5cm/S would track 20cm/S at a force of 2,000 dynes (about 2g).

This is found merely by altering the expression to give velocity equals force divided by impedance. In practice, reasonable accuracy can be achieved by such extrapolation in spite of possible non-linearity occurring in the compliance and resistive elements of the cartridge when the force is changed.

In any case, the threshold of tracking readout is obviously limited in absolute accuracy, no matter how hard one tries to avoid errors; but under good test conditions the error should not exceed about 10%.

## Side-thrust Correction

There is bound to be some error as the result of side-thrust, but at the low-tracking weights employed for low velocity tests this is not significant. It is difficult to determine an absolute correction factor for this, and it is not very easy to apply side-thrust correction when small laboratory weights are used for the tracking weight.

The bearing friction of the arm is also bound to influence the results, but this and side-thrust to some extent are factors found under normal playing conditions; for as yet it is impossible to operate a cartridge without an arm or some means of support and with zero losses and anti-forces!

To tell whether a cartridge is likely faithfully to track a disc recorded to the maximum as provided by the present state of the art demands detailed knowledge of the maximum velocities and amplitudes recorded, and this is not adequately known.

However, the curve in Fig. 4.3 gives some idea of the 'average maximum' recorded levels over the spectrum, and if a cartridge has a sufficiently low tip impedance to handle these at the recommended tracking weight (as based on the tests described), then one can be fairly sure that the cartridge (in a suitable arm with side-thrust correction) will accommodate the majority of music discs without undue flaw.

It is well known, of course, that the RIAA recording characteristic is geared to a reducing bass velocity as a means of preventing the low-frequency amplitude from rising, and there is some evidence that the amplitude in this part of the spectrum is limited to about 0·005cm. The compliance of a cartridge, therefore, must be suitable to handle this degree of amplitude at least without trouble. More details of the RIAA recording characteristic are given in *Pickups and Loudspeakers*. Also see Table 4.1, page 120.

EMI have issued a tracking test record, type TS201, and the curve of this is given in Fig. 4.4. This is similar to the curve in Fig. 4.3, but is based on r.m.s. velocities as distinct from the peak velocities in Fig. 4.3. Peak velocities are 3dB above r.m.s. velocities.

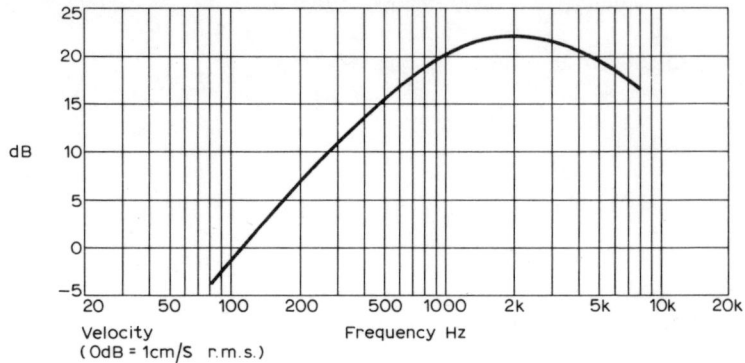

Fig. 4.4 Levels recorded on the EMI TS-201 tracking test disc. Note that these relate to r.m.s. velocity.

With this disc, therefore, it is possible to get some idea of how a cartridge will track at the actual 'average' velocities recorded. The test setup in Fig. 4.1 can be used, the tracking weight then being adjusted until the least waveform mutilation at the recorded frequencies (80Hz to 8kHz) is achieved.

The disc is sweep-frequency recorded, though each side commences with a reference 1kHz band, so it is necessary to adjust the timebase of the oscilloscope to retain the waveform display in usable form over the spectrum. The Y gain will need to be adjusted to keep the amplitude of the display reasonably constant over the swing of velocity.

If the cartridge is unable to track the record at its maximum recommended tracking weight, the tracking ability would be poor. It would be unwise to endeavour to improve the tracking by turning up the tracking weight above the recommended maximum. It is desirable to use optimised side-thrust correction when testing with this disc.

The Haymarket Publishing Group (publishers of *Hi-Fi Sound*) have also issued a test record, type HFS69, which contains 300Hz bands for tracking weight tests in conjunction with side-thrust correction. Again, the setup in Fig. 4.1 can be used to establish the threshold of tracking (or mistracking!).

# Disc Playing Equipment Tests

## Compliance (Static) Tests

The factor of compliance mostly determines the low-frequency tracking ability of a cartridge where the amplitude of the modulation can be large. To measure the static value of this a calibrated shadowgraph is required. The cartridge is mounted so that an image of the stylus is projected on to the screen and a small force is applied to the stylus to obtain a measurable deflection.

The static compliance is then the measured deflection divided by the force required to produce it, and is expressed in terms of $10^{-6}$cm/dyne. Both lateral and vertical compliance should be so measured. A common value of compliance these days is $20 \times 10^{-6}$cm/dyne.

## Compliance (Dynamic) Tests

This is measured by using a low-frequency sinewave recording of known amplitude (a lateral cut for lateral compliance and a vertical cut for vertical compliance) to find the least tracking weight, using the setup in Fig. 4.1 without the high-pass filter if this is likely to interfere with the oscilloscope display, required with optimised side-thrust correction.

The dynamic compliance is obtained by dividing the amplitude by the established force in dynes. For example, if the amplitude is 0·005cm and the force 2,000 dynes, then we have $0·005 \times 10^6/2000$, which works out to $2·5 \times 10^{-6}$cm/dyne.

Some of the CBS discs carry 100Hz bands of peak amplitude 0·001, 0·002, 0·003, 0·004 and 0·005cm to facilitate tests like this, with sections both in the vertical and lateral planes. In effect, though dynamic compliance is a measure of the stylus tip impedance at the particular frequency of test, but expressed in a different form.

The dynamic compliance thus obtained is numerically smaller than the static compliance, and it is noteworthy that the latter is commonly specified by manufacturers.

## Effective Mass of Stylus Tip (Estimating)

The factor of effective tip mass mostly determines the high-frequency tracking ability of a cartridge—the smaller the mass the better the high-frequency tracking. This is because the effective tip mass is accelerated by the modulation, and if the mass is large the acceleration is inhibited (see *Pickups and Loudspeakers*).

It is a very difficult factor to measure accurately, but we can obtain a fair estimate by dividing the force required for tracking a high-frequency band by the modulus of acceleration, which is equal to $2\pi f V$, where $f$ is the modulation frequency and $V$ the velocity.

Thus, if we require a force of, say, 1,470 dynes (equivalent to about 1·5g) just to track a 10kHz sinewave of velocity 10cm/S, we have $1{,}470/6{\cdot}28 \times 10 \times 10^3 \times 10$, which works out to 0·0023g, or 2·3mg. Top-flight cartridges have an effective tip mass below 1mg.

## Mechanical Resistance

This factor cannot easily be measured in isolation since the damping resulting from it affects the tracking performance at low and high frequencies, though mostly at the middle of the spectrum. However, some idea of its effectiveness in damping high-frequency resonances can be gleaned by frequency response, separation and tone-burst tests, yet to be considered.

## Frequency Response Tests

One test setup for measuring the frequency response of a pickup cartridge is given in Fig. 4.5. Here the pickup is correctly loaded to the input of the amplifier, and the output is then taken to an audio millivoltmeter, via a switched attenuator (having 1dB steps). The established tracking weight and side-thrust correction should be used.

Correct loading is necessary to ensure that the response occurs as the manufacturer intended. A magnetic cartridge requires a load of about 50kΩ (47kΩ preferred value is generally suitable), while a piezo cartridge (crystal or ceramic) will require a much higher value load, usually between at least 1MΩ and 4MΩ.

Each stereo channel should be tested in turn, and thorough screening should exist throughout the test circuit to avoid hum induction, and to minimise hum-loop problems it helps if the audio millivoltmeter and amplifier (the former in particular) is battery powered. The channel not being tested should be terminated to a screened load resistor, and the cartridge should be set up in the arm, etc., as already described for the tracking tests.

## Frequency Response Discs

There are numerous discs available for frequency response testing, some recorded to the RIAA standard and others with constant velocity between 500Hz and 20kHz and constant amplitude between

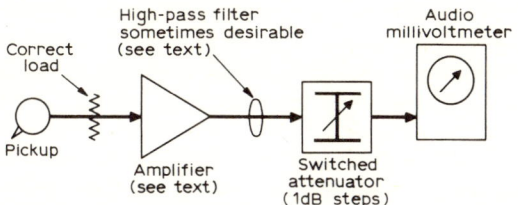

Fig. 4.5 Instrument setup for frequency response and separation measurements (see text).

20Hz and 500Hz. It is essential for the bands of test frequencies to be identified in some way, usually by announcements, and it is less disturbing to work from bands of constant frequency than from gliding tone.

If a disc recorded to the RIAA standard is selected, then the amplifier will have to carry RIAA equalising, which should be accurate over the spectrum to at least $\pm 0.5$dB. However, if the disc has constant velocity modulation over the major part of the spectrum (i.e. from 500Hz to 20kHz) RIAA equalisation should not be used, and in this case, assuming a sufficiently sensitive audio millivoltmeter, the pickup can be loaded, via the attenuator, straight to the instrument, but taking care that the load employed accounts for the resistances of the attenuator and audio millivoltmeter.

In some instances it is necessary to employ a matching pad so that the cartridge sees exactly the resistance to which it should be loaded. This applies particularly when the attenuator/instrument input resistance is lower than the required load value—series resistance then being called for. It can also apply when piezo cartridges are being tested, for although the input resistance of an audio millivoltmeter may be in the region of 1M$\Omega$, this may be insufficient for the cartridge.

Moreover, the input resistance of the amplifier might be much lower than required for a piezo cartridge unless it is designed with bootstrapping or with an FET input stage for high input impedance. The readout will be in serious error unless special attention is paid to these aspects of loading.

## Constant Tone Bands

A suitable RIAA test disc is the EMI TCS-101. This carries bands of constant tones from 30Hz to 20kHz, with each frequency having a band of left channel followed by a band of right channel. Each band is preceded by an announcement of frequency and channel (left or right), recorded laterally so that it will appear in both channels. The 1kHz bands are recorded at 1cm/S r.m.s., and there are also commencing and concluding 1kHz bands of lateral modulation.

The 1kHz left and right bands are best used as the starting reference, and the appropriate one should be played with the pickup tracking at the established tracking weight (or as recommended by the manufacturer) so that the audio millivoltmeter can be adjusted to a suitable readout datum, preferably on the dB scale, using the millivoltmeter controls and the attenuator and ensuring that there is sufficient plus and minus scope on the latter to measure response deviations up and down.

The TCS-101 starts at 20kHz, and from that frequency down to 10kHz the modulation level is cut by 6dB. Thus, when playing the bands down to 10kHz 6dB will need to be taken from the readout circuit by means of the attenuator, but making sure that it is put back again at 10kHz!

The idea is then merely to plot the plus and minus deviations at the attenuator required to maintain the established readout datum at each frequency. When the deviation is less than 1dB interpolation can be adopted at the meter, assuming that it is scaled in decibels.

## High Frequency Accuracy

At the higher frequencies there will be a tendency for the millivoltmeter pointer to waver due to disc ripple and other effects. This can be eliminated by using a high-pass filter in the readout circuit, ensuring that it is switched out when measurements at lower frequencies are made.

Groove/stylus noise, too, can sometimes mask small changes in the high-frequency readout. The noise bandwidth can be reduced by including a low-pass filter in the readout circuit if necessary, but this must not be allowed to affect the in-band measurements up to 20kHz.

The test records must be kept in perfect condition, and frequent

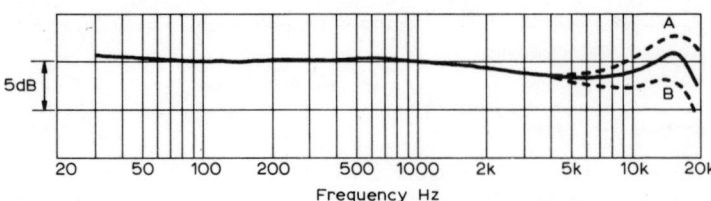

Fig. 4.6 Frequency response of magnetic cartridge. The broken-line curves at the treble end indicate insufficient load A and excessive load B.

replacement is essential if high-frequency readout accuracy is to be maintained. The TCS-101 is recorded identically on both sides, which doubles its life.

**Magnetic Cartridge Response**

The full-line curve in Fig. 4.6 gives some idea of the response to be expected from a good magnetic cartridge of the moving-magnet or induced field variety. The broken-line curves at the treble end show the effects of incorrect loading, with the load too high at A and too low at B. Both channels should be similar.

The mild treble peak could be due to a high-frequency resonance, but in this case it would be desirably damped, indicating well applied and good mechanical resistance. A violent peak, often at a lower frequency, might indicate that the frequency response is extended by utilising a treble resonance, which is the sign of a *poor* pickup cartridge.

The very slight rise at the bass end would possibly indicate the onset of bass resonance, where the effective mass of the arm starts to resonate with the compliance of the cartridge.

**Piezo Cartridge Response**

The response of a piezo cartridge is rarely as 'smooth' as that shown in Fig. 4.6, but a good quality ceramic should fail to exhibit violent changes in response. If such a cartridge has an insufficiently high value load (when tested without RIAA equalisation) there will be early bass roll-off.

However, it is possible to test a piezo cartridge for response via an RIAA-equalised preamplifier. For this the cartridge needs to be loaded across a *low* value resistance which, in conjunction with the capacitance of the piezo element, has the effect of tilting the response so that the treble is lifted and the bass depressed when the disc is recorded to the RIAA standard.

In effect, the cartridge might exhibit velocity characteristics under this condition of loading, meaning that with a constant velocity recording the output remains fairly constant over the spectrum.

The equaliser then corrects this for RIAA recording, as it does with the similar output from magnetic cartridges, but just how 'flat' the resulting curve will be will depend on the design of the cartridge (and its capacitance) and on the nature of its inbuilt equalising.

When a constant velocity disc is used for response measurements, equalisation must not be used, of course, but since most discs of this kind change to constant amplitude below about 500Hz, there will occur a bass roll-off starting at that frequency, so the curve will have to be corrected against the recording characteristic to bring up the response at the bass end; alternatively, any response deviations below 500Hz can be determined with reference to the recording characteristic.

The RIAA characteristic only approximates constant amplitude, resulting from the bass cut and treble lift, and the mean slope of the curve is 4dB/octave. The RIAA recording characteristic is given in Table 4.1 below, and the replay characteristic, of course, is the converse of this.

Table 4.1

| Frequency (Hz) | dB | Frequency (Hz) | dB |
|---|---|---|---|
| 30 | −17·8 | 2000 | 2·6 |
| 50 | −17·0 | 3000 | 4·7 |
| 100 | −13·1 | 5000 | 8·2 |
| 200 | −8·3 | 10000 | 13·7 |
| 300 | −5·6 | 15000 | 17·2 |
| 500 | −2·6 | 20000 | 19·6 |
| 1000 | 0* | | |

Values rounded off to the first decimal place.
* Reference level.

The capacitance across the load, from the screened leads, for example, can also affect the frequency response of magnetic cartridges, and in some cases can incite in-band electrical resonance in conjunction with the inductance of the cartridge signal coils. Tests should thus be made with minimal shunt capacitance, using screened leads of low capacitance.

**Stereo Separation Tests**

The instrument setup in Fig. 4.5 can be adopted for stereo separation tests, and with the two channels of the cartridge correctly loaded, the plan is to establish a readout reference on one channel and then to measure—in decibels down—the crosstalk signal in the other channel, at the correct tracking weight and with side-thrust correction.

The EMI TCS-101 disc is suitable for these measurements because each frequency band commences with the left channel followed by the right channel. Thus, to measure right crosstalk in the left channel, the left channel of the cartridge would be connected to the input of the test setup, and at each frequency a convenient datum would be established on the audio millivoltmeter when the pickup is playing the left channel.

The datum should be obtained in this case by adjusting the audio millivoltmeter for maximum sensitivity and reducing the deflection to the datum by turning on at least 36dB of attenuation.

When the band changes to the right channel, therefore, the readout will be crosstalk signal, so to establish the original readout datum it will be necessary to remove attenuation by operating the switched attenuator. The amount of attenuation removed is a direct measure of the channel separation in decibels.

The converse procedure is adopted to measure left crosstalk in the right channel. The right channel of the cartridge is connected to the input of the test setup, and at each frequency a convenient datum is established on the crosstalk signal when the pickup is playing the left channel. In this case there will be little or no attenuation applied to the readout circuit.

When the band changes to the right channel the readout will be the direct signal, so attenuation will then have to be applied to obtain the established datum, and the amount applied is a measure of left-to-right separation in decibels.

## Separation Limits

Accurate readings much below 30dB are almost impossible to obtain owing to the separation limits of the disc itself and the groove/stylus noise. A high-pass filter is certainly required for these tests when checking at high frequencies to avoid the meter pointer swinging due to disc ripple. etc.

For the most accurate results it is necessary to use a bandpass filter, tunable to each test frequency, so that noise, etc. either side of the test frequency is removed. The filter of a wave analyser can sometimes be used for this purpose.

A typical separation curve (one channel only) is given in Fig. 4.7. The fluctuations at the treble end are often caused by resonances, which may or may not show up on the frequency response curve. The separation is always greatest round the middle of the spectrum, falling off at the bass and treble ends.

## Output Voltage Tests

The output of a cartridge is referred to a specific recorded velocity (level) at 1kHz across the specified value of load, and is given in r.m.s. volts. The velocity chosen is 1cm/S or 5cm/S r.m.s. The test setup is simple, as shown in Fig. 4.8.

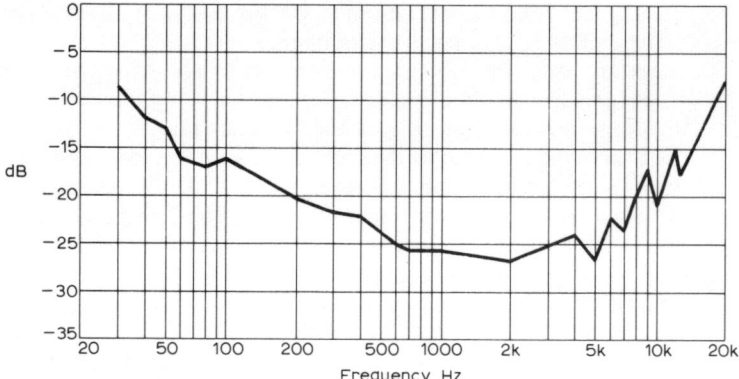

Fig. 4.7 Separation curve of magnetic cartridge. Mild or violent changes in the treble region could indicate resonance.

Care must be taken, however, to ensure that the cartridge sees the correct load resistance, and this will not be the same as the value of the resistor, since this is in parallel with the input resistance (impedance) of the audio millivoltmeter.

Each channel is tested in turn (with the output of the cartridge not being tested connected across a load resistor), and the difference between the two channels of a quality cartridge should not significantly exceed 0·5dB, though there may be greater differences towards the extremes of the spectrum.

The test disc, of course, must be recorded in the appropriate planes (left and right, corresponding to 45/45-degree cuts). The 1kHz 1cm/S bands on the EMI TCS-101 disc are suitable, as also are the left and right bands on the EMI TCS-102 disc, which differs from the former in that it is gliding tone and has the left channel recorded on one side and the right channel on the other side.

The HFS69 disc also has 1kHz reference bands, left and right, recorded at a level of 5cm/S. The correct tracking weight and side-thrust correction should be used for this test.

On lateral modulation each channel of a cartridge should give an output $1/\sqrt{2}$ times that obtained from left or right modulation at the same recorded level and frequency, with minimal imbalance.

### Tone-Burst and Squarewave Tests

The damping and hence the transient performance of a cartridge can be appraised to some extent by playing a disc carrying squarewave or toneburst modulation and viewing the output on an oscilloscope. The cartridge should be correctly loaded and adjusted

## Disc Playing Equipment Tests 123

Fig. 4.8 (left) Instrument setup for measuring pickup output voltage.

Fig. 4.9 (right) Instrument setup for displaying transient performance and 'ringing'.

for tracking weight and side-thrust correction and fed to the Y input of the oscilloscope, as shown in Fig. 4.9. There is not generally any need for filtering; in fact, filtering could change the shape of the wave displays.

The HFS69 disc features tone-burst bands and examples of displays from this are given in Fig. 4.10. (a) shows 16 cycles of burst followed by a 16-cycle period of suppressed tone. (b) shows 2 cycles of burst followed by a 2-cycle period of suppressed tone,

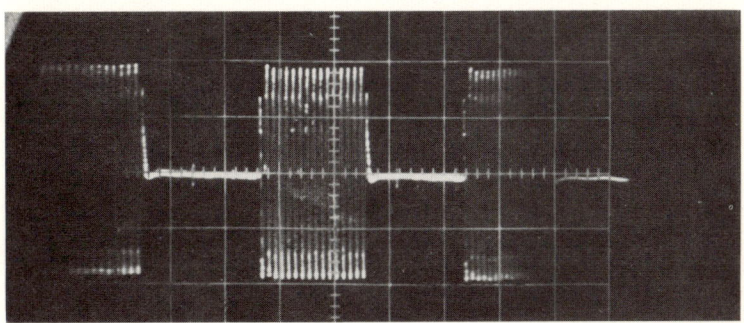

Fig. 4.10 Pulsed-tone oscillograms: (a) above, (b) below.

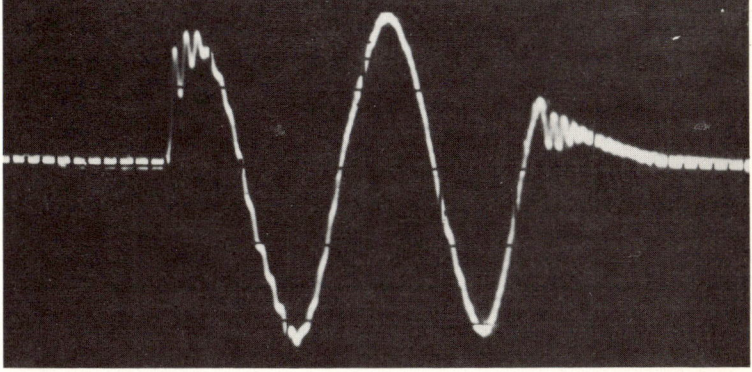

while (c) and (d) also show 2-cycle bursts, but with (d) apparently carrying 'rings' of approximately twice the frequency as those carried by (c).

Actually, (c) was obtained with the disc spinning at $33\frac{1}{3}$ r.p.m., while (d) was obtained with the disc velocity approximately doubled. The timebase of the oscilloscope was adjusted to give the same time period at both speeds. Based on time, therefore, the 'rings' at (d)—in spite of being expanded along the X axis—are at a similar frequency to those at (c).

This proves that the 'rings' are in fact resulting from the cartridge and are not implanted in the groove of the test disc. Obviously, if they were implanted they would have the same ratio to the sine-wave display on both oscillograms.

The 'ring' frequency can give some idea of the high-frequency resonance, while the amplitude indicates the effectiveness of the internal damping. For example, the 'rings' at (b) have a greater amplitude than those at (c), which could mean that the cartridge responsible for (c) has better damping than that responsible for (b). Nevertheless, such displays are not always so meaningful and should thus be interpreted with care.

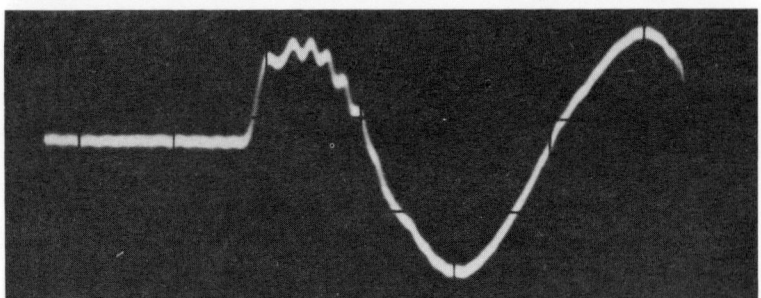

Fig. 4.10   Pulsed-tone oscillograms: (c) above, (d) below.

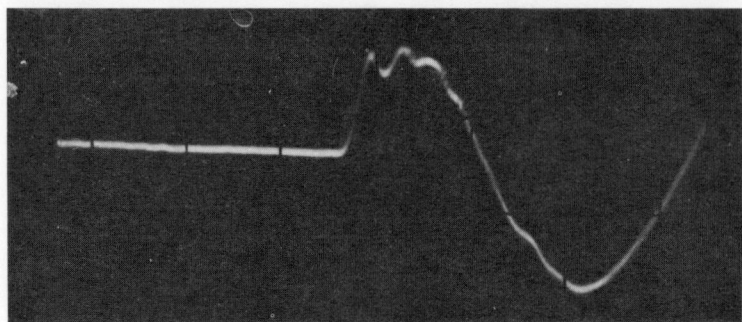

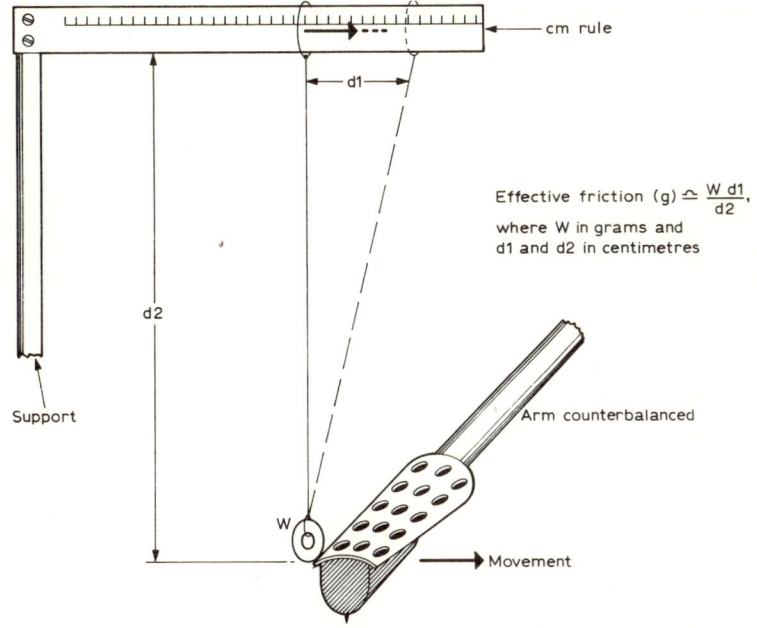

Fig. 4.11 A method of measuring lateral friction of a pickup arm.

The oscillogram at (a) shows how rapidly the cartridge responds to the onset and conclusion of the burst, without undue 'ringing', thereby indicating good transient performance.

## PICKUP ARM TESTS

The five main parameters of a pickup arm are (i) bearing friction (both vertical and lateral), (ii) side-thrust correction force, (iii) effective mass at the head, (iv) tracking weight adjustment, including counterbalancing and (v) tracking error. Let us deal with these in turn, starting with bearing friction.

### Bearing Friction Tests

The friction effect in which we are interested is that reflected to the head in terms of a force required to overcome the friction and thus instigate movement. Force gauges, capable of measuring in any direction, are available, but since these rarely read down to a few tens of milligrams it is necessary to find the force required to produce movement at a measured distance close to the pivot, and then calculate the effective force at the head end of the arm in terms of the ratio of the distances involved (see Fig. 4.12).

However, the lateral friction can be measured quite accurately by a simple exercise in mechanics, shown in Fig. 4.11. I use a small tripod carrying a centimetre rule and a 500mg or 1g drilled weight tied at the bottom of the thread. The longer the thread the more accurate the measurement. By using about 90cm of thread and a 5g weight quite accurate readouts are feasible.

The idea is to balance the arm for zero tracking weight and hang the weight so that it is just touching the headshell in line with the stylus of the cartridge, and then slide the thread along the ruler until the arm is just set into motion.

The 'starting friction' is approximately equal to the weight in grams times the displacement of the thread along the rule in cm divided by the length of the cord, including the weight to its point of contact with the headshell, in cm. Since the force required for the start of motion in the opposition direction may be slightly different, it is best to make a further measurement in that direction and find the mean of the two.

For obvious reasons the vertical friction cannot be measured in this manner. However, a fair assessment of the force required to break this friction can be obtained by balancing the arm for zero tracking weight as before, and then placing milligram laboratory weights on top of the headshell, over the stylus, until movement just occurs.

If milligram weights are not available, a larger weight can be placed at some point along the arm, close to the pivot, and the effective force at the stylus required just to break the friction can be calculated as shown in Fig. 4.12. The same calculation is employed when a force gauge is used along the arm towards the pivot.

**Side-Thrust Correction Force Tests**

Either a force gauge or the setup in Fig. 4.11 can be used to measure the side-thrust correction force at any point over the swing of the arm and at any force setting on the arm. Most arms provide for a reducing force as the arm swings from the outside to the inside groove diameter, and the 'average' change is about 10 to 15%. Some arms, though, exhibit a substantially constant force over the whole swing, while few are designed so that the force increases as the cartridge traverses the record.

Side-thrust results from the head offset, such that the frictional drag of the stylus in the recorded groove produces a force displaced from the normal axis of the arm which is reflected as a torque at the arm bearing, causing the arm to be subjected to a force bias towards the centre of the disc.

The resulting unbalance of lateral forces at the stylus means that the stylus is pressing harder on the inner wall of the groove than on the outer wall. Not only can this disturb the 'stereo balance' of the cartridge, causing a greater crosstalk in one channel with respect to that in the other, but the extra lateral force needs to be outweighed by a greater tracking weight for a given tracking performance.

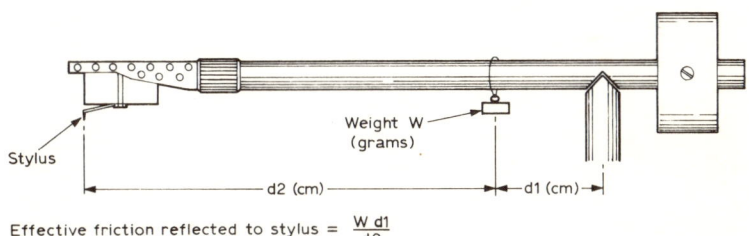

Effective friction reflected to stylus = $\frac{W \, d1}{d2}$

Fig. 4.12 When milligram weights are not available, a larger weight can be placed along the arm, as shown, and the force required just to break the vertical friction can be calculated from the expression given. The weight is best suspended on thin cotton and slid along the arm as required.

Since the side-thrust is a function of the velocity of modulation, tracking weight, co-efficient of friction between the stylus and the moving groove and the dimensions of the stylus tip, absolute correction is impossible under normal playing conditions, and it is not perfectly clear how the side-thrust changes with reducing groove diameter.

Nevertheless, the application of some correction is beneficial when the arm has low bearing friction, since it can reduce the required tracking weight by at least 10 per cent for a given tracking performance and reduce unbalance effects between the two channels.

The HFS69 test disc carries various 300Hz bands at different modulation levels with lateral and vertical cuts for the determination of side-thrust adjustment and minimum tracking weight. The photograph of Fig. 4.13 shows the side-thrust correction force being measured towards the middle of the arm swing with a Correx force gauge.

## Effective Mass Tests

The mass of the arm remains even when it is counterbalanced to zero with the cartridge *in situ*. It is this effective mass which is seen by the compliance at the stylus, and which resonates with it to produce a rising output at low frequencies.

Since a cartridge has two modes of compliance—vertical and lateral—it follows that there could be two bass resonances. However, a good arm usually succeeds in damping these so that the rise in bass output is minimal.

Nevertheless, if the resonance happens to coincide with motor rumble frequency, loudspeaker bass resonance or some other resonance, including that of the room, unwanted noises in the sub-bass region can be encouraged, as also can acoustic feedback, with the loop completed in part acoustically from the loudspeakers to the record playing equipment and in part electronically, via the amplifier.

If the resonance is very low—say, below 10Hz—unstable tracking can result. The resonance is equal to $1/2\pi\sqrt{(MC)}$, where $M$ is the effective mass of the arm in grams and $C$ the compliance of the cartridge in c.us. Thus, if $M$ is about 10g and $C$ about $10 \times 10^{-6}$cm/dyne the resonance will occur around 15Hz.

Fig. 4.13 Showing the measurement of side-thrust correction force using a Correx gauge. A dial gauge for measuring the play in the turntable bearing is shown at the right.

Fig. 4.14 The author's scheme for measuring effective arm mass (see text).

Since the slip frequency of the drive motor often occurs at 22·5Hz the bass resonance should fall below or above this frequency, but not much below 10Hz for the reason already stated. When above, the resulting rapid bass roll-off helps to delete the rumble.

Some of the CBS test discs carry bands gliding to a very low-frequency so that the bass resonance of a pickup system can be observed by a rise in cartridge output, and if the frequency and cartridge compliance are known the effective mass of the arm can easily be calculated.

## Measuring Arm Mass

A scheme that I have devised for measuring arm mass is pictured in Fig. 4.14. Here the arm is counterbalanced to zero with a Goldring 850 cartridge in the headshell. The stylus assembly is slipped out to expose the magnet underneath the main body, and this is lined-up with the axis of a ferrite rod which passes through the air core of an inductor.

Also positioned in the core, with freedom of movement, is a coiled spring of non-magnetic metal and the top of this is clamped to the headshell, The setup is carefully adjusted so that the arm will vibrate vertically with about 5mm clearance between the cartridge magnet and the top of the ferrite rod.

The inductor is then energised with audio sinewave from an oscillator direct or via an amplifier and the frequency adjusted until the vertical compliance of the spring resonates with the effective mass of the arm, which is indicated by vigorous vertical vibration of the arm.

The cartridge is removed from the head-shell and secured by itself to the top of the spring, adjusting as previously described, and laboratory weights are placed on top of the cartridge and adjusted in value until resonance occurs again. The total value of the weights is a close approximation of the effective mass of the arm.

Tests can be made with and without the headshell, with the counterweights at various positions and over the range of tracking force.

I have found that the Wharfedale 4mH inductor, produced some time ago for loudspeaker crossover filters, is ideal for this purpose. A wood block at the bottom secures and locates the ferrite rod and coiled spring.

### Tracking Weight and Counterbalancing Tests

When an arm carries a calibrated tracking weight adjustment the accuracy of this can be measured with a simple balance and laboratory weights or with a force gauge. It is also possible to obtain a good idea of the accuracy by counterbalancing the arm to zero with the tracking weight adjustment at maximum, then reducing the tracking weight in 1 gram intervals, while adding 1 gram laboratory weights to the headshell to discover whether true balance is restored each time.

With the headshell *in situ*, the counterbalancing should cater for a wide range of cartridge weights. Most arms will conterbalance a cartridge weighing down to 3 grams (indeed, some go down to zero gram) and up to tens of grams. The range of weight acceptance can be measured by placing laboratory weights on top of the headshell when the cartridge is removed.

### Tracking Error (Lateral)

Another important arm parameter is the lateral tracking error. This results because the groove of a record is cut with the recording head traversing a true radial path while the cartridge on replay follows a curved path because, of course, the arm is pivoted at the far end. The angle of the plane of vibration of the stylus thus varies relative to the groove, and the larger this angle, the greater the lateral tracking error.

The longer the arm, the smaller the error: but long arms have the disadvantage of additional mass, and the improvement in tracking performance is not significant merely by length increase of a few inches. The problem is minimised by offsetting the angle of the head on the arm and by arranging the geometry so that the stylus tip overhangs the centre of the disc by a predetermined amount.

However, the audio technician rarely has call to measure or calculate the tracking error, but it is noteworthy that the distortion arising from lateral tracking error increases with decrease in linear groove velocity and with increase in recorded velocity. Thus, the distortion is the least at the outside diameter of the record where the velocity of the groove relative to the stylus is greatest, increasing as the groove plays out.

The distortion is essentially second harmonic, and its value (percentage) is equal to $1 \cdot 75 \phi V_r / V_g$. where $\phi$ is the tracking error in radians, $V_r$ the recorded velocity and $V_g$ the groove velocity. The distortion resulting from a straight arm (no overhang or offset) of about 20cm in length can rise to some 7% at 10cm/S, since the lateral tracking error could range from 22 degrees at the outer radius and 7·5 degrees at the inner radius.

An arm of the same length with correctly applied overhang and offset will reduce the error to a maximum of about 3 degrees, and the distortion to about 2%. It is usual to adjust the overhang so that the tracking error is almost zero at the inner radius. An alignment protractor can be used for this adjustment. as explained in *Pickups and Loudspeakers*.

## Tracking Error (Vertical)

Modern discs are cut with a 15-degree vertical tracking angle and the least distortion on replay occurs when the stylus of the cartridge is designed with a similar angle. It is thus important for the cartridge to be accurately set-up in the headshell and the arm on the motor board (especially in terms of height), usually so that the top surface of the cartridge is parallel with the surface of the disc on the turntable.

It is not possible accurately to determine the vertical tracking angle (or error) from physical measurements. One procedure is to measure the second harmonic distortion produced by the test cartridge when it is playing a disc recorded with bands of 400Hz sinewave with vertical cuts ranging from 6 degrees of trail to 43 degrees of rake.

The band which produces the least distortion is that with the nearest cut to the cartridge's vertical tracking angle. A disc carrying such bands is the CBS STR160.

### Pickup Distortion Tests

It is not particularly easy to measure pickup distortion when the turntable has even very mild shortcomings in terms of wow and flutter. However, the basic procedure is as already outlined in the preceding chapters, although third-octave filters are commonly employed for the measurements, and these can be used to determine not only second-harmonic distortion, but also third- and fourth-harmonic when the pickup is playing suitable sinewave test records of low recorded distortion (at about 5cm/S level).

Modern cartridges possess minimal inherent non-linearity, and the distortion resulting from them is mostly due to the geometric shortcomings of the recording and replay functions, such as those resulting in tracking error, etc.

The second-harmonic distortion from a quality cartridge should not exceed about 5% at any frequency at a recorded level of 5cm/S in the appropriate plane when accurately set up. The distortion, though, can rise significantly in excess of this when the tracking errors are above nominal, especially at high recorded velocities.

Sometimes a test of IMD is made from special records carrying 400 or 200Hz and 4kHz tones over a range of recorded velocities up to 18dB above $1 \cdot 12 \times 10^{-3}$cm peak amplitude.

## TURNTABLE UNIT TESTS

The four main parameters of a turntable unit are (i) rumble, (ii) wow and flutter, (iii) speed stability and (v) hum induction, and to appraise these the three components of disc playing equipment (namely, the pickup cartridge, the arm and the turntable unit) have to be employed collectively. Let us deal with these in turn, starting with rumble.

### Rumble Test Set-up

If we want to determine the rumble produced by a specific turntable unit, then this has to be adequately mounted into a test rig containing motor board, pickup arm and cartridge. The arm and cartridge

## Disc Playing Equipment Tests

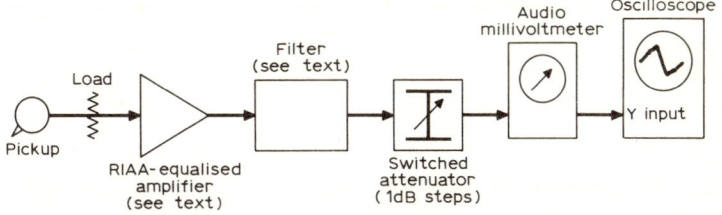

Fig. 4.15 Instrument setup for measuring rumble.

combination should possess a known low-frequency resonance characteristic with the well damped peak occurring below 22·5Hz (the motor's slip frequency). The stylus should be new and have a well polished tip to minimise groove noise, and the motor board should be very solid and non-resonant.

When a complete record playing unit is tested all the components as fitted should be adopted, and no special precautions need to be taken to minimise plinth and motor board resonances.

A suitable test setup is given in Fig. 4.15. The cartridge is wired for mono (i.e. for lateral modulation) by connecting the two channels in parallel and correctly loaded to the input of an amplifier. Some tests require this to be equalised to RIAA, in which case the magnetic pickup preamplifier section of a hi-fi amplifier could be used (taking the output from the tape recording socket), or one of the small self-contained magnetic pickup preamplifiers made by such people as Shure, Goldring, etc.

When the amplifier is not equalised a weighting filter can be employed (see Fig. 1.2) which rolls off the treble to remove the noise and also the bass to give a readout related more closely to the annoyance value of the rumble. Filtering may also be used when the amplifier is equalised, and it will be appreciated, of course, that the readout will be influenced by the nature of the filtering, etc., which should be stated.

### Rumble Test Routine

The idea is to establish a readout datum on the audio millivoltmeter with plenty of attenuation switched in when the pickup is playing a suitable band of 1kHz sinewave modulation of lateral cut. The pickup should then be allowed to track unmodulated grooves and the attenuator switched down until the readout reference datum is obtained by the meter responding to rumble signal and possibly noise.

The ratio of noise to rumble can be measured if necessary on the oscilloscope connected to the output of the audio millivoltmeter, and the amount of attenuation switched down in decibels is a direct measure of rumble and noise.

The level of the reference 1kHz modulation will affect the ratio, so this should also be stated. Common modulation levels are 5 and 10cm/S. The HFS69 test disc, for example, carries a 1kHz 10cm/S lateral band followed by unmodulated grooves for this test.

The photograph in Fig. 4.16 shows a rumble test in action. The audio millivoltmeter behind the record playing unit is reading mostly rumble signal, which is being displayed by the oscilloscope on the right. Below the oscilloscope is the RIAA-equalised amplifier and behind this the switched attenuator and several filters used in the test.

It is possible to measure the main frequency components of the rumble from the X axis in terms of time on the oscilloscope, and in Fig. 4.16 these worked out to about 20Hz.

**Wow and Flutter Tests**

Wow is the result of cyclic speed variations below about 20Hz, while flutter is the result above 20Hz, up to about 200Hz. Tests are usually made with a wow and flutter meter, such as the Rank-

Fig. 4.16   A rumble test underway (see text).

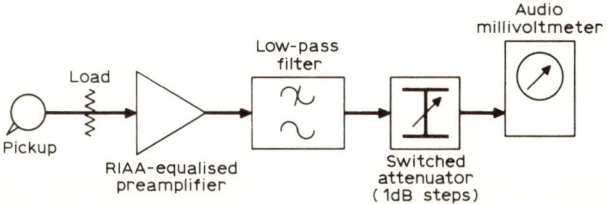

Fig. 4.17 Instrument setup for measuring hum induction from turntable unit to pickup cartridge (see text).

Studio Flutter Meter Type 1740, when the pickup is playing 1kHz or 3kHz sinewave modulation from a lateral cut of 5cm/S. A filter is often used between the pickup and the wow and flutter meter, and the readout is either unweighted or weighted percentage deviations from the mean frequency in peak or r.m.s. values.

For example, percentage flutter is equal to $f_{max} - f_{min} \times 100/f_{av}$, where $f_{max}$ is the highest deviation from the average frequency, $f_{min}$ the lowest deviation from the average frequency and $f_{av}$ the average frequency.

High grade turntables yield less than 0.1% r.m.s. wow and flutter, which is below the subjective threshold of about 0.3% wow and 0.15% flutter, both r.m.s. values. The readout, however, is significantly affected even by a mildly off-centre disc, so the test disc must possess a high standard of concentricity and should fit tightly on to the spindle of the turntable.

The HFS69 test disc has a 3kHz wow and flutter band with lateral cut to 5cm/S and a concentric locked groove cut at the rim for establishing the concentricity before the actual test is made.

### Speed Stability Tests

Speed stability refers to the 'drift' in speed over the long term. This can be tested with a stroboscope, while a wow and flutter meter, using the discriminator tuning control, can be used to give the percentage error.

Tests should be made at different mains voltages and with different loads on the spinning disc, such as would be imposed by high velocity modulation and by the use of a groove cleaning device (e.g. 'Dust Bug').

### Hum Indication Tests

Since the hum induced from a turntable motor to a pickup cartridge is a function of the susceptibility of the cartridge to hum pick up, it is possible to perform only comparative tests based on a specified cartridge. A suggested test setup is given in Fig. 4.17.

The cartridge is correctly loaded (both channels in parallel) to a preamplifier carrying RIAA equalisation, and the output is passed through a low-pass filter (to remove noise, with the $-3$dB point at about 500Hz and the rate of roll-off 12dB/octave) and switched attenuator to an audio millivoltmeter. If possible, both the preamplifier and audio millivoltmeter should be battery-powered.

All attenuation is switched out and the audio millivoltmeter adjusted to a high sensitivity to read the residual hum and noise when the motor is disconnected from the mains supply. A suitable readout datum is thus established on the audio millivoltmeter.

The mains supply is then connected to the motor and the arm carrying the cartridge moved across the surface of the turntable, until a maximum reading is obtained, this possibly demanding increased readout circuit attenuation. The amount of attenuation switched in to secure the original reference is a direct measure in decibels of the hum induction relative to the particular cartridge employed. The test can be repeated if necessary on the left and right channels only to simulate stereo-playing conditions.

**Other Turntable Unit Tests**

The movement or play in the turntable bearing as reflected to a specific turntable diameter can be measured with a dial gauge, such as can be seen to the right of Fig. 4.14.

Other tests include the time that it takes a turntable to stop from a given velocity (usually $33\frac{1}{3}$ r.p.m.), the efficiency of the lowering/lifting device, etc.

CHAPTER FIVE
# SYSTEM TESTS

A 'SYSTEM' CAN BE REGARDED either as a connected assembly of separate items or as an integration of two or more items which, although complementary, differ in character. For example, a hi-fi *system* could be composed of an amplifier, pair of loudspeakers, radio tuner and record player.

A tuner-amplifier is really a 'system' because it combines an amplifier and radio tuner in an integrated form. By the same token, a radio receiver is a complete system because it integrates the tuner and audio sections with a loudspeaker.

The previous chapters have concentrated on the tests of separate items of equipment, and while it is possible to perform such tests relative to the separate sections of an integrated system, it is often more convenient and meaningful to test two or more sections together.

**The Tuner–Amplifier**

A good example is the tuner–amplifier (sometimes called 'hi-fi receiver'). The tuner section of this can be tested separately as described in Chapter 3 by taking the output prior to the amplifier from the detector or stereo decoder (or, perhaps, more conveniently from the recording signal socket of the amplifier), while the amplifier section is tested as detailed in Chapter 2.

The integration may be in advance of that of the tuner–amplifier, and may include the record playing equipment. In this event it is sometimes desirable to test certain parameters of the record playing section *in situ*, via the integration, rather than testing in terms of isolated items. Integrated tests of this kind would be for hum induction, acoustic feedback, cartridge tracking, equalisation, etc.

While loudspeakers are incorporated in radio receivers and some radiograms, the type of 'quality' audio equipment towards which this book is focused, rarely—if ever—adopts integration to this extent. The loudspeakers are invariably separate items of the system to reduce acoustic feedback problems and to permit advantageous placement from the stereo listening and room resonance aspects. These are considered in some detail in the complementary books—*Tuners and Amplifiers* and *Pickups and Loudspeakers*.

## TUNER–AMPLIFIER TESTS

Because the amplifier department of a tuner–amplifier incorporates preamplifiers and inputs for source signals other than those from the integrated tuner section, the amplifier is best tested as a separate item—as though there were no tuner integration. The tests detailed in Chapter 2 are thus applicable to the amplifier section.

As already mentioned, the tuner section can also be tested in isolation, the tests detailed in Chapter 3 then being applicable. However, since the integrated tuner will always be used with the integrated amplifier, there is some merit in testing the tuner in conjunction with the amplifier, and I now propose to discuss tests of this nature.

### Hum and Noise Tests

The S/N ratio of a tuner is referred to 30% or 100% modulation, but when an amplifier is connected to the output of the tuner there is also the hum and noise produced by this to be taken into account, and since the level of these unwanted signals is significantly affected by the setting of the volume control, we must first establish a setting for this based on the depth of test modulation.

Personally, I prefer referring the overall hum and noise to 100% modulation and setting the volume control for the rated output power when the modulation is 400Hz or 1kHz. A suggested test setup is given in Fig. 5.1.

The f.m. generator is correctly matched to the aerial input of the tuner–amplifier (using an artificial aerial, balun or matching pad of some kind when necessary, as explained in Chapter 3) and the amplifier output is terminated with a resistive load of the rated value and wattage (Chapter 2). The load is connected to an audio millivoltmeter via a switched attenuator when this is not included on the audio millivoltmeter.

With the volume control of the tuner–amplifier turned well down, the readout attenuation is adjusted to determine that the r.f. input is of a sufficient level to take the tuner section well into amplitude limiting. This, of course, is indicated by the output remaining constant when the r.f. signal from the generator is increased.

The tuner–amplifier and test equipment should be allowed fully to reach a stable operating temperature, after which the tuner should be accurately adjusted to the generator signal.

The plan then is to advance the volume control until the rated power is delivered to the load, as can be calculated by measuring the audio voltage, squaring it and dividing the result by the load resistance in ohms. For 100% modulation, maximum power is commonly delivered when the volume control is set approximately to the 1 o'clock position, which corresponds closely to the setting established on music signal for maximum dynamic range without the peaks clipping, and is the control setting adopted for the remainder of the test.

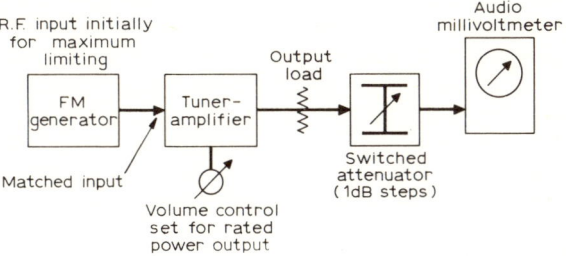

Fig. 5.1  Setup for hum and noise tests of tuner-amplifier.

The next move is to adjust the audio millivoltmeter and attenuator for a convenient readout datum at maximum power, ensuring that at least 60dB attenuation is switched in. After this the modulation is switched off and the output level of the generator adjusted from zero upwards in convenient steps, noting at each step the amount of attenuation switched down from that used at the rated power to provide the readout datum. The results can be plotted to produce a curve as shown at A in Fig. 5.2. This, of course, has much in common with the S/N ratio curve of f.m. tuners.

A curve revealing the limiting performance can also be plotted, as shown at B in Fig. 5.2. This is obtained with the modulation running in the same way as explained in Chapter 3, but referring zero dB to maximum *power* at the modulation level used.

It must be remembered that curve A indicates the amount of hum and noise below the rated output of the amplifier section when the volume control is set in the manner explained. A different result would be expected by checking the S/N ratio of the tuner section alone, since filters might be used in such a test and the hum and noise of the amplifier section would not be included.

However, in my judgement, it is the overall performance of a tuner–amplifier which is the most meaningful, particularly when the the volume control is set as for maximum music signal drive.

Some tuner–amplifiers incorporate a preset between the tuner output and the amplifier input to facilitate equalising the level of the radio signal. In the absence of specific setting up instructions, this is best adjusted on 100% modulation so that maximum power is obtained with the volume control set between 1 and 2 o'clock.

### Receiver Tests and Output Power

Sensitivity and hum and noise tests can be made in the a.m. section of a tuner–amplifier, and the procedures adopted would be similar to those expressed in Chapter 3, based on a power amplifier readout as just discussed. In fact, the basic tests normally adopted for a.m. radio receivers would be applicable in many cases. The output would not then normally refer to maximum or rated power.

For receiver tests a *standard output power* is often used, which is 50mW—i.e. 17dB(mW)—or, when there is a particular reason for a higher power, 500mW might be used, corresponding to 27dB(mW). If a lower power would be more suitable for the test, then 7dB(mW) or 5mW is a value often chosen. The chosen standard output, however, must be specified.

The *maximum useful output power* is related to a specified percentage of distortion, while the *reference output power* is 10dB below the maximum useful output power. The power is r.m.s.-based and is generally measured at a frequency of 1kHz.

### IHF Procedure

IHF recommend that the volume control be adjusted for 20dB attenuation when the tuner section of a tuner–amplifier is being tested. At this setting most tuner–amplifiers yield the rated output, or marginally less, when the r.f. signal is modulated to 100%. Clearly, if this setting results in audio overdrive on 100% modulation the subsequent test results would be meaningless, which is why I prefer to set the volume control initially for the rated output on 100% modulation.

IHF tuner tests are made in the recommended manner (Chapter 3), but through the power amplifier into the rated load resistance when the filters and controls affecting the frequency response are switched

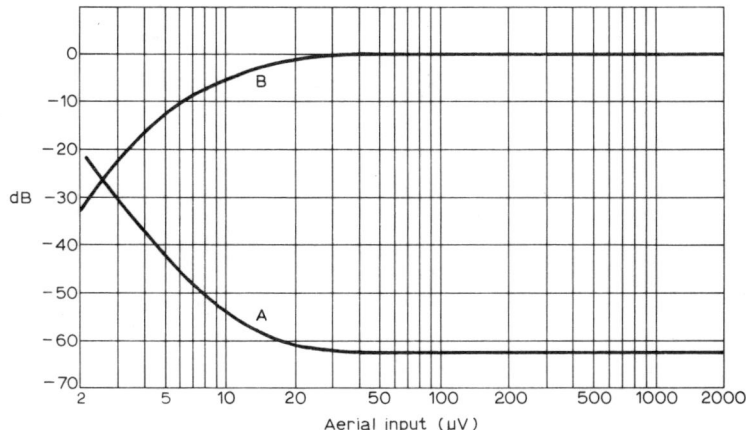

Fig. 5.2 Limiting (B) and hum and noise (A) curves appropriate to a quality tuner-amplifier. 0dB here corresponds to the rated output obtained with the volume control suitably adjusted on 100% modulation.

out and adjusted as indicated by fascia markings for the flattest response.

Except when making measurements relating specifically to stereo operation, only one audio channel of a stereo tuner–amplifier needs to be involved. The left-hand one is generally the most convenient from the switching point of view. However, when making tests in the audio department it is often desirable to compare the performance of one channel with that of the other in some if not all the parameters.

**Radio Breakthrough**

Owing to the high $f_T$ of the transistors used in contemporary solid-state audio equipment radio breakthrough is not uncommon when a system switched to gramophone pickup is operated in a strong radio signal field. In locations close to a powerful radio or television transmitter the breakthrough may occur when sources other than pickup are switched. The pickup source is the most vulnerable because it is the most sensitive and most easily overloaded.

The radio signals are detected by the resulting front-end non-linearity when the active devices are pushed by the strong signals towards overload. The high gain at this point in the system lifts the level of the diminutive audio signals sufficiently to cause a disturbing background interference on low-level disc material.

A.M. television sound signals can also be similarly troublesome, while the vision signals are responsible for a disquieting buzz, the magnitude of which alters with changing picture content.

While it is usually the preamplifier stages which are responsible for the effect, the final stages of the control section and the power amplifier cannot be ruled out completely. Indeed, a recently developed tuner–amplifier was found to be virtually unusable after dark on any source owing to unwanted reception of a multiplicity of short- and medium-wave stations, and this occurred even with the volume control turned right down!

The trouble developed after dark because then the signal field rises from medium-frequency stations at distant sites due to ionospheric propagation. All stations are received together, of course, because there is no tuning, the amplifier input circuit being aperiodic right up to the $f_T$ of the transistors or the roll-off frequency of any filters.

Fortunately, trouble like this can usually be cleared or significantly reduced by the addition of low-pass filters to the inputs of the affected stages. When the breakthrough continues with the volume control turned right down, as just mentioned, the pick-up obviously occurs somewhere after the volume control, unless the control is feedback operated.

However, one can easily locate the entry area by 'shorting' the input of each stage in turn to chassis using a capacitor of some 2nF. A direct short should be avoided as this would affect the transistor biasing conditions when d.c. couplings are used.

### Radio Breakthrough Susceptibility Test

Clearly some form of test is required to measure the degree of susceptibility of the tuner–amplifier or amplifier to modulated radio signals and to determine the efficiency of any filtering subsequently incorporated. Such a test has been devised and the instrument setup required is shown in Fig. 5.3.

The test procedure is as follows. With the amplifier appropriately switched and loaded, a signal corresponding to the modulation frequency of the a.m. signal generator (400Hz or 1kHz) is applied from the audio oscillator to the input being tested. The tone controls are set 'flat', the filters switched out and the volume control of the amplifier turned to maximum.

The level of the audio oscillator signal is then increased until the voltage across the output load corresponds to the rated power of the amplifier. This can be measured by setting the switched attenuator to 0dB and using the audio millivoltmeter direct. It is desirable

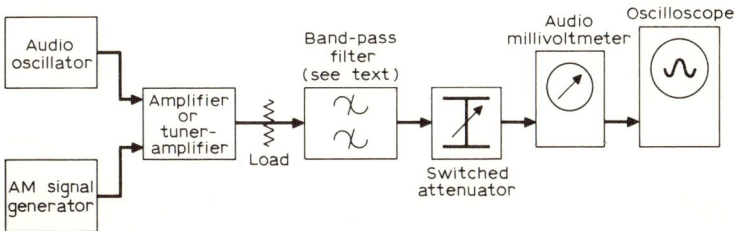

Fig. 5.3 Setup for testing the susceptibility of an amplifier or system to radio breakthrough.

to monitor the signal waveform on an oscilloscope to ensure that it is free from clipping, etc.

At least 70dB attenuation is then switched into the readout circuit, and the sensitivity of the audio millivoltmeter correspondingly increased until a convenient reference datum is established.

The audio oscillator is then disconnected from the input and the a.m. signal generator connected in place. As already noted, the modulation frequency should correspond to the frequency of the audio signal previously applied and, of course, the bandpass filter should also be tuned to that frequency. If a bandpass filter is not available, a 12dB/octave high-pass filter can be used instead, the −3dB point of which should occur at about 200Hz, with a low-pass filter if necessary to delete the r.f. signal. The modulation should be set to a depth of 30%.

Now, the level of r.f. signal applied will depend on the sensitivity of the input being tested. Suggested levels are 5mV on magnetic pickup inputs and other similar low-level inputs and 100mV on all higher level inputs.

The readout attenuation is then switched down until the audio millivoltmeter and oscilloscope indicate the presence of the modulation of the r.f. signal when the frequency is set initially to about 100kHz. At this stage, the a.m. generator should be tuned over the various radio and television bands, making sure that the r.f. output remains constant, until the frequency providing the largest readout is discovered. This frequency should be noted, as also should the frequency range over which the readout responds.

The generator should then be tuned to the frequency of maximum readout and the switched attenuator adjusted to provide the originally established reference datum. The amount of attenuation switched out corresponds, in decibels, to the radio breakthrough susceptibility of the equipment under test. Both the frequency of maximum response and the range of frequencies over which a response is obtained should be stated in the test results.

Many contemporary amplifiers respond to radio signals well into Band III (the television band), though there is a tendency for a mild peak to occur at about 5MHz; but this can depend on the nature of the amplifier's circuits. Desirably, the breakthrough should be below noise, but it is not uncommon to obtain values of a mere 35dB at the magnetic pickup and similar low-level inputs and 55dB at the other high-level inputs.

Small wonder, then, that so many amplifiers and tuner–amplifiers of recent design are suffering the problems of severe radio and television breakthrough when operated close to powerful transmitters!

**Radio Filtering**

Some manufacturers are taking heed of the problem and incorporating suitable filters at the inputs of the vulnerable stages. These are not complicated devices and in general can consist of simple RC networks at the base input of the early stages, as shown in Fig. 5.4.

The components concerned are C and R. The resistor R is merely connected in series with the input circuit, while C is connected close to the transistor. directly between base and emitter, using the shortest possible leadout wires.

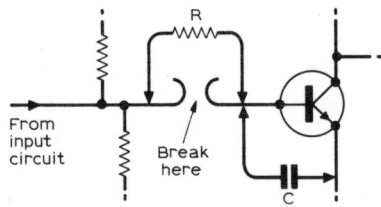

Fig. 5.4   The connection of a simple RC filter for reducing radio breakthrough.

Values are not critical, and in general R can be about 2·2kΩ (though some circuits may require a smaller value) and C around 100pF. A too high a value for C, of course, will affect the treble performance of the stage.

The input transistors of each stereo channel will need to be so processed, and while in some cases only the pickup preamplifier input transistor will require attention, in others subsequent stages will also need attention. However, the inclusion of the base/emitter capacitor only (between 47 and 100pF) is sometimes all that is required.

## System Faults Leading to Breakthrough

The trouble can be aggravated by ineffective screening of the input circuits of the system, by too long leads and by leads of a critical length (corresponding to a half wavelength of the interfering signals). The pick up of the radio (or television) signals is either on the source connecting leads or on the actual components and circuits of the equipment itself. However, the latter is minimised by efficient screening, which is generally a design factor.

The signal can also be picked up by the cartridge or arm, and some cartridges appear to be more vulnerable in this respect than others.

There are also times when the r.f. signals gain admittance to the amplifier section from the loudspeaker cables, via the negative feedback loop, the cables then acting as aerials! In severe cases, therefore, it sometimes helps to screen these, and to 'earth' the braiding to the common amplifier or equipment earth point which, itself, should desirably be bonded to a true earth.

However, it should be noted that excessive reactance in the loudspeaker circuits is undesirable, so the screened cable should be of low capacitance and employed only when absolutely essential.

## Impulsive Interference

Another problem is mains-borne impulsive interference of the kind that is responsible for the staccato crashes which seemingly put the loudspeakers in peril each time a thermostatically controlled item of domestic electrical equipment comes into operation. Such equipment containing heavy reactance, like the electric motors of refrigerators with capacitive phase control and solenoid switching, etc., create most disturbance by injecting transient pulses into the mains circuit.

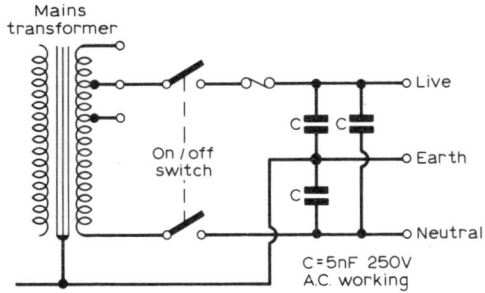

Fig. 5.5  Delta capacitor filter at the power input of an amplifier or system.

If the audio system is connected to the same mains circuit—like a ring mains—the pulses can get into the amplifier circuits, via the power supply, and activate the loudspeakers as just described.

Simple radio-type suppression at the offending item of electrical equipment is generally totally inadequate, and more drastic filtering at or in the system is required, and even then severe pulses can rarely be completely eliminated.

Delta capacitor suppression at the mains input (see Fig. 5.5) can sometimes help; but in general filtering of the amplifier power supply lines is required in addition. Fig. 5.6 outlines the general scheme of capacitor filtering at the output and input of the power rectifier, where C are the suppression capacitors of value around 5nF each.

This multiplicity of capacitors may not be required, though it might be necessary to include similar value components at each supply feed line, particularly on the lines powering the preamplifier.

It is noteworthy that amplifiers with a series transistor regulator are less prone to impulsive interference, of the mains-borne variety, than counterparts without such regulation.

At the time of writing a method of testing the vulnerability of amplifiers to mains-borne impulsive interference is under development by the author, but the essential problem lies in devising a device to deliver pulses to the mains input of the amplifier in a controlled manner to simulate the mains supply circuit.

However, it would appear that amplifiers with good radio breakthrough immunity are less troubled with impulsive mains-borne interference than those which are wide open to r.f. signals, so the filtering in Fig. 5.4 could help.

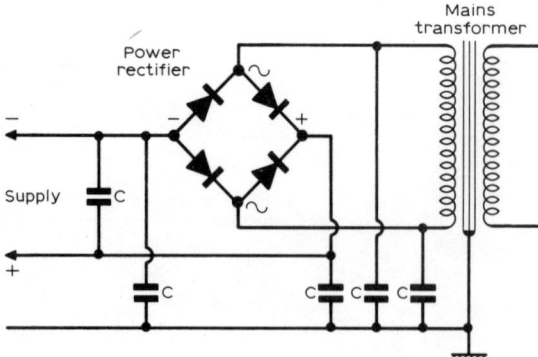

Fig. 5.6 Power supply filtering can help to reduce the system response to impulsive mains-borne interference. The filter capacitors are marked C in the diagram and should be suitably rated with values of about 5nF.

## Channel Balance Tests

The overall electrical balance between two stereo channels of an audio system can be determined by using the set-up shown in Fig. 5.7. Here each channel is correctly loaded at its output and an audio millivoltmeter is connected across the 'live' terminals of the left and right channels. A mono in-phase signal is then applied to the inputs so that each channel is driven together.

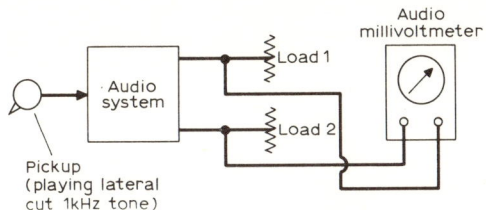

Fig. 5.7 System channel balance test setup. The audio millivoltmeter should preferably be battery-powered or, if mains-powered, should not be earthed.

When the two channels are in perfect electrical balance only residual hum and noise will be indicated by the millivoltmeter. Thus, the balance control should be adjusted for a readout null, which corresponds to the setting for electrical balance relative to the input programme signal source.

When the source is a gramophone pickup the lateral 1kHz bands on the EMI TCS101 test record can be employed, the amplifier then being switched for stereo operation. This test will take into account unbalance effects in the pickup cartridge and the readout will indicate noise, etc. of disc replay due to vertical movements of the stylus producing anti-phase signals.

When the amplifier is switched for mono operation the two stereo channels are effectively connected in parallel at the input so the anti-phase signals at the pickup will be cancelled. A null will still be obtained as before, but will fail to take into account unbalance of the source signals.

Sources other than disc, of course, can be applied, including tape replay and stereo radio when modulation of suitable phase is recorded on to the two stereo tracks of tape or modulated on to the v.h.f. carrier. It is possible to use the stereo test tones transmitted by the BBC (page 99) to check the stereo balance through the system from the stereo decoder.

The stereo balance from the listener's point of view must take into account the loudspeakers, their positioning and the acoustics of the listening room, so setting the balance control as described for the electrical balance may well not correspond to the acoustical balance.

To test this one needs a sound level meter (or similar device) placed midway between the two loudspeakers or at the stereo listening position, as shown in Fig. 5.8. A suitable test tone is applied to one channel and the audio millivoltmeter adjusted to provide a convenient readout datum. The balance control should then be adjusted until this same reading is obtained when the test tone is applied to the other channel.

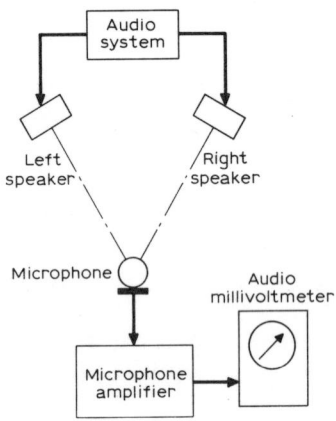

Fig. 5.8 Setup required for testing the overall channel balance, taking into account the loudspeakers and room acoustics, etc.

It is instructive to check at various frequencies over the spectrum to see how the balance is maintained at the extremes as well as at the middle of the audio range. Loudspeaker and listening room shortcomings tend to upset the balance at low and high frequencies due respectively to bass resonances and treble absorption.

Reflections, too, can play havoc with the acoustic balance, so it is often necessary to establish a compromise balance control setting, based on tests at low, middle and high frequencies.

### Stereo Identicality Factor (BSI)

This is a test which provides meaningful information on the balance of an amplifier or system over the important part of the audio passband. It is expressed as the ratio of the mean output voltage of the combined channels when both of them are fed with equal in-phase signals, to the mean output voltage of the combined channels

## System Tests

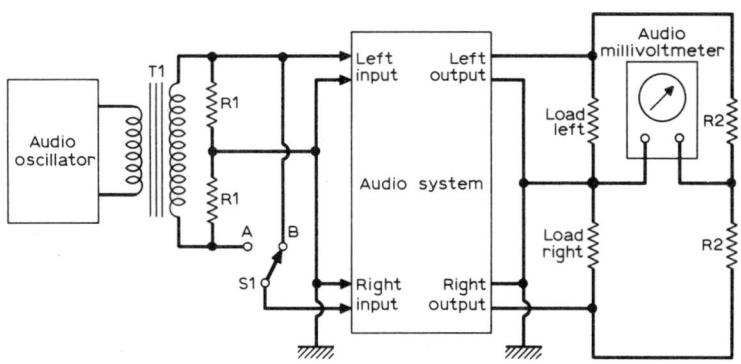

Fig. 5.9  Instrument setup for testing stereo identicality factor (B.S.I.).

when both of them are fed with equal anti-phase signals. The test setup is given in Fig. 5.9.

Both channels are loaded at the output in the usual way and the system is adjusted for 'flat' operation with the volume control at maximum and the balance control at mid-position (i.e. for optimum balance).

T1 is a coupling transformer receiving the audio oscillator signal at its primary and delivering it, relative to a centre-tap provided by the balancing resistors R1, at its secondary. R1 are equal value resistors, of course, and have a value to match the source. SW1 operates such that at position A the inputs receive anti-phase signals and at position B in-phase signals.

Balancing resistors (of equal value—about 100-ohm) are also fitted across the 'live' terminals of the loads at the output so that the audio millivoltmeter indicates the mean value of the combined output.

SW1 is set for antiphase input signals and the audio oscillator is adjusted for 1kHz and to provide a signal level to drive each amplifier approximately to the reference output power, which is 10dB below the maximum useful output power of the amplifiers, see page 36.

The sensitivity of the audio millivoltmeter is increased sufficiently to provide a usable readout and the balance control adjusted for minimum indication while progressively increasing the sensitivity of the readout. SW1 is then set for in-phase input signals.

Tests are made from low frequencies up to at least 3kHz, and the ratio of voltages resulting from the in-phase and anti-phase inputs is expressed as a dB value at each test frequency. The results can be plotted to provide a convenient curve, such as that shown in Fig. 5.10.

150                  *Audio Technicians Bench Manual*

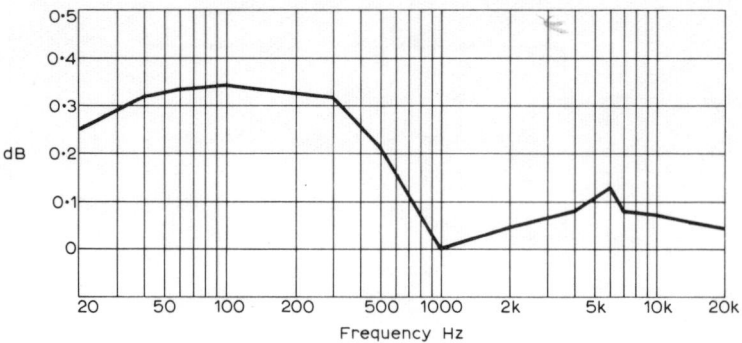

Fig. 5.10  Spectral balance curve (see text).

It is desirable to repeat the tests at lower settings of the volume control, for when ganged controls are used it is not uncommon for the balance to diminish significantly at low settings. The degree of balance control adjustment required to restore balance should be specified.

This test takes into account the phase characteristics of the amplifiers over the frequency spectrum. However, it is possible to appraise the phase performance of each channel separately by means of an oscilloscope, now to be explained.

**Phase Tests**

The test setup is given in Fig. 5.11. The audio signal is applied to the input under test, and the output is loaded in the usual manner. The output across the load is connected to the Y input of the oscilloscope, the timebase is switched off and the input signal direct applied to the X terminal of the oscilloscope. If the oscilloscope is without an X amplifier, then an attenuator will be required in the Y signal feed so that equal deflection of the test signal can be obtained horizontally and vertically.

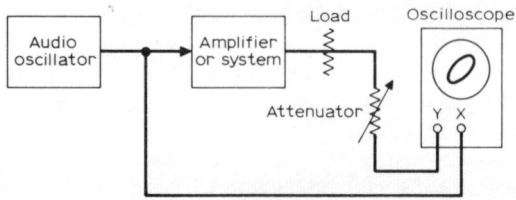

Fig. 5.11  Setup for the display of phase shift on an oscilloscope.

## System Tests

A straight diagonal line is obtained when there is zero phase shift, while a 180-degree shift is indicated by the line altering to the complementary diagonal. When the X and Y deflections are exactly equal and the phase shift 90 degrees the display will be a perfect circle. Intermediate values of phase shift are indicated by ellipses, as shown in Fig. 5.12.

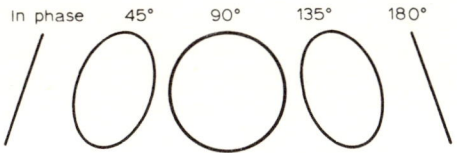

Fig. 5.12  Phase shift patterns.

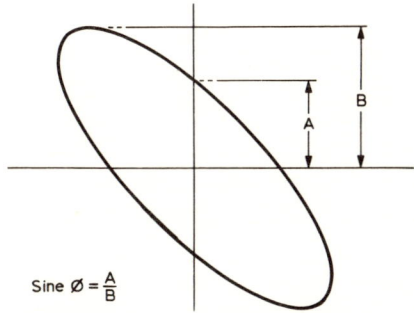

Fig. 5.13  Showing how the angle of phase shift can be calculated.

The phase shift can be roughly calculated from the display as shown in Fig. 5.13. Clipping is shown by flattening at the ends of the ellipse, while distortion is shown by kinking of the ellipse or diagonal line.

### System Crosstalk Test

Stereo crosstalk in amplifiers, tuners and pickups is examined in Chapters 2, 3 and 4. The important factor, however, is the overall system crosstalk. Obviously, the stereo separation can be no better than that of the item in the system responsible for the most crosstalk. Thus, if the pickup cartridge, for example, has mid-spectrum separation of 20dB, then the overall separation cannot be better than this, even though the amplifier separation may be as high as 50dB.

For good stereo reproduction the mid-spectrum crosstalk should not be worse than 20dB (26dB is a better value), while at 100Hz and 10kHz the separation should still be in the region of 10dB (16dB is a better value here).

Clearly, when one stereo item of a system is connected to another the inter-channel signal leakage paths are effectively in parallel. This is rather the same as resistors being in parallel. Thus, if one path results in a crosstalk ratio of 100:1 (40dB) and another path 10:1 (20dB), then the separation will be slightly less than 20dB. In fact, the reciprocal of the total ratio is equal to the sum of the reciprocals of the other ratios involved, so in the above case the total ratio works out to a little over 9:1, corresponding to about 19dB.

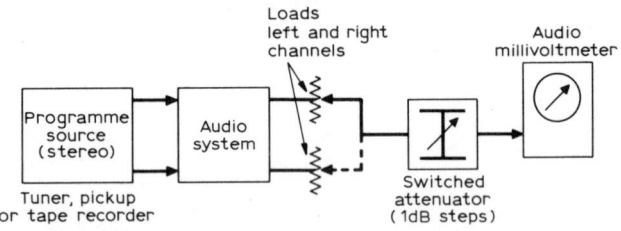

Fig. 5.14    Setup suitable for measuring system crosstalk.

The setup for measuring system crosstalk is given in Fig. 5.14. With the programme source connected and producing signal in one channel only, the audio millivoltmeter is connected across the load of the speaking channel and adjusted with the switched attenuator (the latter having at least 50dB of attenuation switched in) for a convenient reference datum.

The audio millivoltmeter is then connected across the load of the non-speaking channel and the attenuation switched down until the reference datum is obtained. The amount of attenuation switched down is a direct measure of crosstalk or channel separation in decibels.

Tests should be made at various settings of the volume and tone controls, for the latter in particular can often alter the separation ratio when adjusted.

The programme sources should carry suitable test tone (preferably at 1kHz to establish the mid-spectrum separation, and at lower and higher frequencies to check the separation at the spectrum extremes). For example one of the EMI test records would be suitable for testing from a gramophone pickup, a test tape from a tape recorder and a stereo signal generator or the BBC's test tones from a stereo radio tuner.

## System Frequency Response Test

The setup given in Fig. 5.14 can also be used to check the system frequency response. Here one channel is tested at a time and the programme source must carry test tones (either fixed-frequency bands or gliding tone) over the entire audio spectrum from about 20 or 30Hz to 20kHz.

The attenuator and audio millivoltmeter are adjusted to establish a suitable reference readout at 1kHz, and the response plotted in plus or minus decibels relative to this at each frequency, using the switched attenuator to obtain the dB deviations.

This method of using a switched attenuator is more accurate than trying to determine the deviations relative to an established datum on the decibel scale of the audio millivoltmeter. However, for rapid checks the dB scale readout method is often more convenient and is, in fact, practised extensively for more exacting tests by some technicians.

As intimated earlier in this book, some audio millivoltmeters incorporate accurate attenuators, and when this is the case it may not be necessary to employ a separate attenuator for the various tests detailed in this and the foregoing chapters. If the audio millivoltmeter attenuators switch over, say 6, 10 and 20dB, as some do, then the intermediate dB values can be read off the dB scale.

## System Distortion Test

THD or IMD can be measured from the output of a system by using the setup given in Fig. 5.15. The programme source must, of course, carry single-frequency sinewave tone of low distortion for THD measurements, while for IMD tests the source must carry two tones of suitable frequencies and level ratio.

The method of measurement is the same as that described for amplifiers in Chapter 2 (see page 50). However, the harmonic distortion will be significantly greater from the output of a system because the distortion produced by the source will be added in quadrature to that produced by the system.

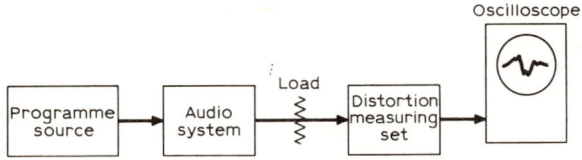

Fig. 5.15  Setup for measuring system distortion.

Quite high levels of distortion result from tape and disc replay (that is, relative to the very low distortion contribution of a hi-fi amplifier, for example), and even some top quality equipment might exhibit disc replay distortion (at 1kHz and 5cm/S) of several per cent, increasing with recording velocity!

**System Hum Level Test**

The hum and low-frequency noise produced by a system can be measured with the setup given in Fig. 5.16. The source is caused to yield 1kHz sinewave tone and, with SW1/SW2 at position A, the volume control is advanced until the reference power (10dB below the rated power) occurs at the load.

This is measured with the audio millivoltmeter reading the r.m.s. voltage across the load direct, such that $W = V^2/R$, where $V$ is the r.m.s. voltage and $R$ the load in ohms.

At least 60dB of attenuation is switched in and the sensitivity of the audio millivoltmeter advanced to secure a convenient readout datum. SW1/SW2 are then switched to position B, and the null filter carefully adjusted until the 1kHz tone is deleted (the null filter can be the notch filter of a THD test set).

The attenuator is next switched down until the millivoltmeter again reads the reference datum. The amount of attenuation in decibels switched down is a measure of the hum and low-frequency noise, including that of the programme source. The hum and low-frequency noise below the *rated power* is 10dB greater than the value indicated by the attenuator.

The high-pass filter is used to remove the high-frequency noises, and the rate of roll-off should be at least 12dB octave with the $-3$dB point round 2·5kHz. The 1kHz band-pass filter clears hum and noise from the reference test signal.

The test can be performed without the null filter by removing the 1kHz tone from the programme source after establishing the reference datum. However, when the source is a record player this should be tracking an unmodulated groove and when a tape recorder unmodulated tape should be passing the replay head. With a stereo tuner the pilot tone should be active even though the modulation is switched off. A tuner, of course, should also be accurately adjusted to the carrier frequency and the carrier should remain when the 1kHz modulation is removed.

The test thus takes into account hum and low-frequency disturbances throughout the whole system, including modulation hum, disc and tape rumble, stereo decoder low-frequency noises, etc.

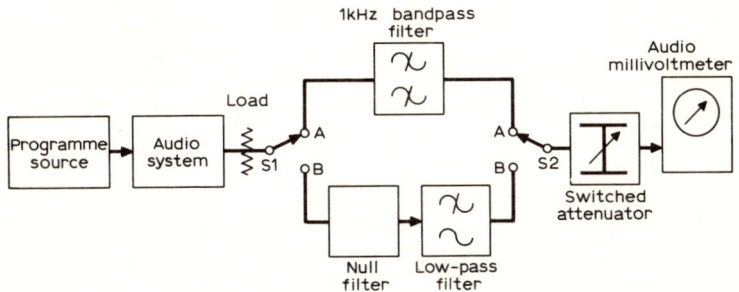

Fig. 5.16  Setup suitable for measuring system hum level, etc.

## Acoustic Feedback Test

The setup in Fig. 5.17 can be used for acoustic feedback tests, particularly when the programme source is a record player.

The loudspeaker is disconnected from the system's amplifier and replaced by a suitable load resistor, the 'live' side of which is coupled, via a calibrated attenuator, to the input of an auxiliary amplifier which is endowed with a uniform frequency response and negligible phase shift over the entire audio spectrum, and whose output impedance is adjusted to correspond to that of the system's amplifier.

The system's loudspeaker is then connected to the output of the auxiliary amplifier. The attenuator setting required for unity gain between the load and the loudspeaker is established as a reference, and about 10dB of attenuation is turned on initially.

Next, the volume control of the system's amplifier is fully advanced and the tone controls and filters set to 'flat'. Then, with the pickup placed on the outer groove of a record, with the turntable stationary, the auxiliary circuit attenuation is reduced until acoustic feedback

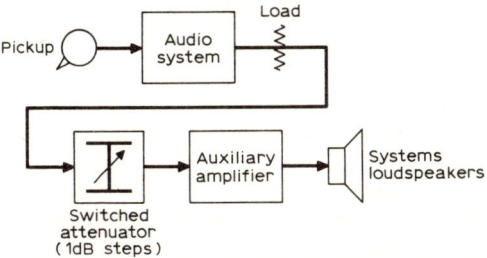

Fig. 5.17  Setup (after B.S.I.) for determining system acoustic feedback.

just occurs. The attenuator is then backed off slightly until the feedback ceases.

The setting of the attenuator is noted in terms of gain or loss in the auxiliary channel, relative to the previously established unity gain reference. The amount of loss is a measure of the acoustic feedback and the amount of gain a measure of the feedback stability reserve of the system, both expressed in decibels.

A stereo system is best tested with a dual-channel auxiliary amplifier and with both channels and loudspeakers of the system in circuit simultaneously, using the stereo pickup of the system, operating in the stereo mode.

The tests can be repeated with discs of unmodulated grooves being tracked by the pickup when the turntable is in motion.

**Loudspeaker Impedance Test**

The impedance of a loudspeaker system can be tested with the setup in Fig. 5.18. The plan is to adjust the audio oscillator initially to 400Hz, corresponding to the frequency at which the impedance of loudspeaker systems is commonly specified, and with SW1 set to position A adjust the signal level for about 250mV on the audio millivoltmeter.

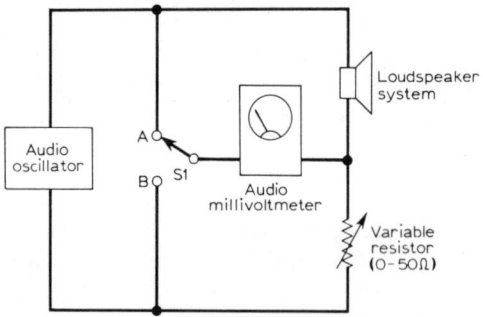

Fig. 5.18  Setup for measuring loudspeaker system impedance.

SW1 is next switched between positions A and B and the variable resistor adjusted for the same reading. The value of the variable resistor is then measured with an accurate ohmmeter, the reading obtained corresponding to the impedance of the loudspeaker system at the test frequency.

The test should be repeated at a number of frequencies from 20Hz to 20kHz so that an impedance curve can be plotted as shown in Fig. 5.19. The rise in impedance at the bass end is caused by system

## System Tests

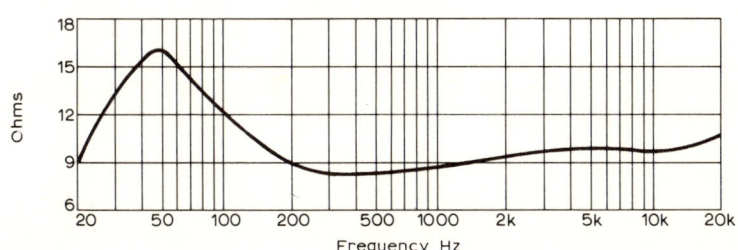

Fig. 5.19   Loudspeaker system impedance curve.

resonance (see *Pickups and Loudspeakers*), while the mild treble rise is caused by inductive reactance.

### System Input Impedance Test

It is often required to know accurately the input impedance of an audio system (amplifier, tuner–amplifier, record player, tape recorder, etc.) at a specific frequency or over a range of frequencies. The setup in Fig. 5.20 can be employed for making tests of this kind.

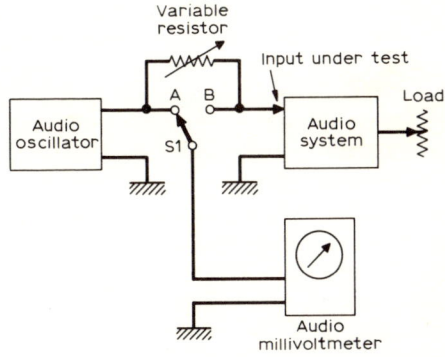

Fig. 5.20   Setup for measuring system input impedance.

The idea is to set SW1 to position A, load the amplifier and set all controls for normal operation. Initially the test is made at 1kHz, and the level of the oscillator signal is adjusted to a value round the nominal sensitivity of the input under test. The audio millivoltmeter is then accurately adjusted to measure this signal, choosing a range to provide as near as possible to full-scale deflection.

With SW1 at position B, the variable resistor is adjusted until the audio millivoltmeter reads exactly half the original value. The input impedance is given by measuring the value of the resistance when removed from the circuit. Clearly, the variable resistance must have a range to correspond to the expected input impedance.

The effect of input reactance can be determined by measuring at other frequencies over the spectrum; but when measuring equalised inputs care must be taken to avoid amplifier overdrive as the frequency is reduced. Tests can also be made to determine how the input impedance is affected by the various system controls.

**System Tone Controls and Filters**

For tests of this kind the setup is similar to that required for basic frequency response tests, as shown in Fig. 5.21. In fact, the response at 1kHz with the tone controls 'flat' and filters out is first plotted to provide a reference datum.

The signal from the audio oscillator should remain at a constant level over the spectrum (monitoring may be necessary), and the level should be set to cater for response lift and cut without the danger of the system running into overdrive or noise respectively.

When measuring the response of the filters and the tone controls in their 'cut' positions a 1kHz 0dB datum should be established on the audio millivoltmeter and the switched attenuator must be adjusted so that at least 20dB attenuation can be removed from the readout circuit at the extremes of the spectrum. Conversely, when measuring the response of the tone controls in their 'lift' positions and the response of the loudness control the attenuator should have at least 20dB available to be switched in.

The plan is to adjust the attenuator at each test frequency to obtain the established 0dB datum and then note the number of decibels switched in or removed at each frequency so that suitable curves can be plotted. Fig. 5.22 shows a number of such curves, while Fig. 23 shows the response of the tone controls, with the broken-line curve showing the reference frequency response with the tone controls 'flat'.

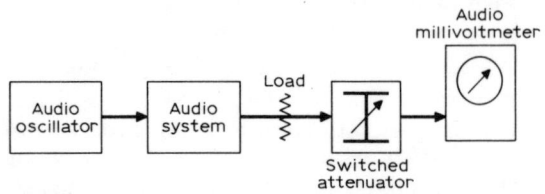

Fig. 5.21 Setup for measuring the frequency response of filters and tone controls.

*System Tests* 159

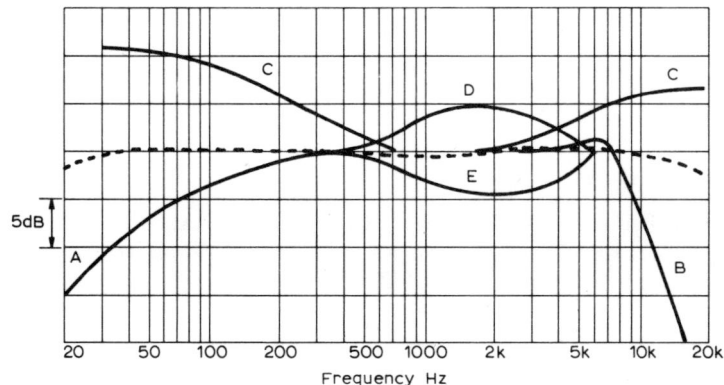

Fig. 5.22 Filter responses relative to the frequency response shown by broken line. A high-pass (rumble) filter; B low-pass (scratch) filter; C loudness control; D and E mid-range filters.

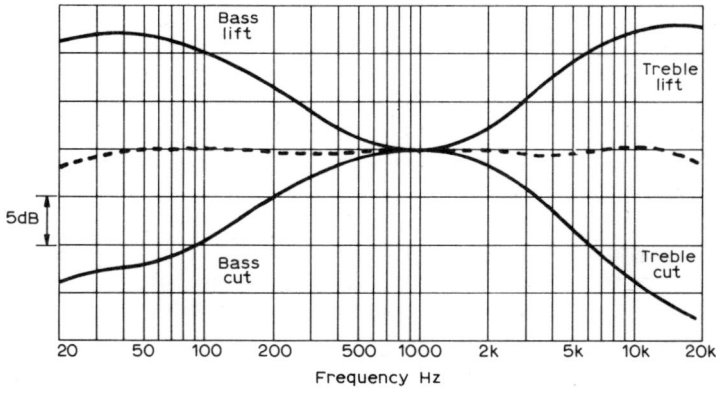

Fig. 5.23 Tone control response curves at maximum and minimum settings of the controls, where the reference frequency response is shown in broken line. Curves at intermediate settings of the controls are sometimes useful.

## TAPE RECORDER TESTS

Like a record player, for example, the tape recorder can be regarded as a 'system' and is thus correctly placed in this chapter. Many of the tests already considered are equally applicable to the tape recorder, including those of frequency response, equalisation, power output, distortion, crosstalk, etc., and require no further description.

Readers desirous of a close study of the subject are recommended to H. W. Hellyer's book entitled *Tape Recorders*, which is in the 'How to Choose and Use series by the publishers of this book. It delves into basic principles, servicing procedures and tests.

## Distortion Tests

While THD and IMD are tested by the procedures already explained, it must be understood that since a tape recorder embodies the actual function of recording the distortion resulting from this should also be included in an *overall distortion test*. It is thus necessary to make a recording of test tone from a low distortion audio oscillator at a specified recording level and then to use the tape so recorded as the signal source for the distortion measurements on replay.

The recording level, the nature of the tape, the degree of distortion on the bias signal and its amplitude and frequency, the tape-to-head velocity, the degree of non-linearity in the recording and replay amplifiers, etc., can all influence the overall distortion, which is commonly in the order of several per cent even on some of the best domestic equipment.

The tape is best recorded at the level specified by the manufacturer, based on the recording level indicator. If this is a VU meter, then 0dB makes a good recording level datum, though it is useful to see how the distortion falls at lower recording levels and rises at higher ones. Another recording level is 32mM/mm.

It is also useful and instructive to plot the distortion over the spectrum at a fixed bias signal level and at a fixed frequency (say, 1kHz) and at different bias signal levels, both at different speeds. A family of curves at different frequencies, with the frequency as the parameter, plotting bias current against THD, can be extremely revealing!

## Measuring Bias Current

The simplest way of measuring the bias current in the recording head is shown in Fig. 5.24. Here the 'earthy' side of the winding is disconnected and a 100Ω resistor included in series. The signal voltage across this is then measured with a sensitive audio millivoltmeter whose frequency response extends to the bias frequency. The bias current is equal to the measured voltage divided by 100.

A similar procedure can be adopted to measure the erase head current, while the distortion on both the bias and erase signal can be measured with a THD setup.

It is possible, of course, to measure the distortion produced separately by the recording and replay amplifiers, and from comparative measurements a fair assessment of the distortion yielded by the recording and replay functions can be gleaned.

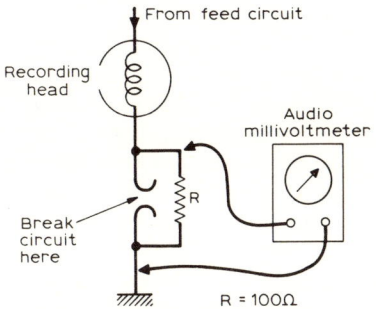

Fig. 5.24 Method of testing record head bias current. A similar procedure can be used for measuring erase head current.

## IMD Tests

IMD tests tend to highlight the performance of tape recorders more than THD tests, and while the latter requires a single low-distortion oscillator for recording the tape, the former requires two oscillators of frequencies and level ratio demanded by the particular type of test involved (see Chapters 1 and 2).

The signal provided by an IMD measuring set should be applied to the input of the tape recording amplifier and a recording made of the signal. The IMD is then measured in the usual manner when the recorded tape is producing the replay test signal.

## S/N Ratio

The *overall* signal-to-noise ratio of a tape recorder should take into account the noise produced by the recording amplifier, the tape and the replay amplifier. Since the S/N ratio is a function of the level of the recording, this should be specified in the test results. On replay, however, the noise is commonly referred to maximum output, as in S/N ratio tests of amplifiers (see Chapter 2).

The reference frequency is 1kHz, and a tape recorded with sine-wave signal of that frequency and at a specified level is used to establish the readout datum on replay. The tape is then run through the recorder set to the record function, but with the volume control fully retarded, so that the recording is erased.

With the originally established replay control settings, the erased tape is once more run through the machine, but this time in the replay mode and the readout attenuation adjusted to provide the previously established reference reading on the audio millivoltmeter. The procedure is similar to that adopted for S/N ratio tests of amplifiers, explained in Chapter 2.

A filter can be used in the readout circuit to remove the low-frequency noises and hum; but it is common for the ratio to include hum and noise of all types.

Greater accuracy is achieved by using a 1kHz bandpass filter in the readout circuit when establishing the reference datum by playing the 1kHz tape recording. Without such a filter, of course, the ratio is not true S/N but S + N/N.

The test results are expressed in decibels below the rated power of the replay amplifier and at a specified recording level. A desirable aspect of this particular test is that it includes noise resulting from imperfect erasure and from distortion on the erase signal.

The amount of noise due to the erasing function can be assessed by running the replay part of the test (when the noise is being measured) either with brand new tape or tape that has been erased by a bulk eraser.

It is also possible to measure the noise of the replay amplifier alone and compare the result with the noise produced when brand new or erased tape is running through the machine to obtain an idea of the degree of noise produced by the tape replay function.

Test tapes are available for measuring distortion, S/N ratio, etc. These are recorded to specified levels. However, in general a recording level of 0dB (on a VU meter) is a good reference for S/N ratio tests; another level is 20dB below peak recording level; but for the results to be meaningful full information of the test conditions must be stated.

### Frequency Response Tests

Test tapes are also available to appraise the replay equalisation. Measurement is based on a frequency response test, using a switched attenuator and audio millivoltmeter in the readout circuit.

The overall 'equalising', taking into account any pre-emphasis provided by the recording amplifier, is best checked by recording a test tape with a wide range of frequencies over the audio spectrum, using the machine which is to provide the replay readout. The test tape is then merely run through the machine in the replay mode and the response deviations at the various test frequencies plotted in decibels.

It is noteworthy that some tape recorders are designed with

significant treble recording pre-emphasis as an artifice for holding the response up at the treble end on replay. This can lead to high overall distortion at the treble end of the spectrum, particularly at high recording levels.

### Wow and Flutter Tests

For an accurate appraisal of wow and flutter a special wow and flutter test-set is required. The test procedure is similar to that described on page 134 in Chapter 4 for turntable units.

The wow and flutter instrument analyses 3kHz (or thereabouts) test tape signal in terms of deviations from the average frequency and gives a readout in peak or r.m.s. values. A peak value can be converted to a r.m.s. value by multiplying by $1/\sqrt{2}$, while a r.m.s. value can be converted to a peak value by multiplying by $\sqrt{2}$.

The instrument contains an input preamplifier and amplitude limiter channel followed by a form of discriminator, and the output from this, which is a measure of the frequency deviations, is fed either direct or through a weighting filter to the readout device. This can be either digital or analogue, the former as a 'counter' and the latter as a directly reading meter or pen-graph indicator.

## ACOUSTICAL CHARACTERISTICS

An audio system containing a loudspeaker can only be measured overall by taking into account the loudspeaker part of it. The same also applies to a loudspeaker system, which itself might be connected to an amplifier, tuner–amplifier, tape recorder, record reproducer, etc.

### Acoustical Frequency Response

The setup for this measurement is given in Fig. 5.25. The audio system is fed with 1kHz signal from the audio generator, applied so

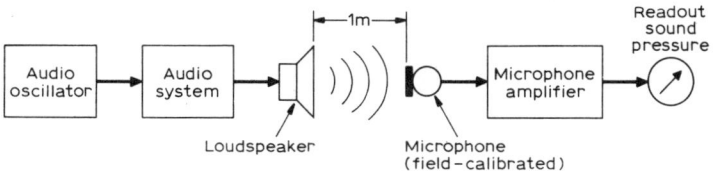

Fig. 5.25 Setup for appraising the frequency response of an audio system or loudspeaker system based on sound pressure.

that a constant e.m.f. is passed through the specified load impedance and, with the volume control at maximum, filters out and tone controls 'flat', the input level is adjusted for a standard output (50mW or at other stated level—page 140) or lower level should overloading of the audio system or loudspeaker occur.

A field-calibrated microphone (that is, one which registers the sound pressure that would exist where located if it and its consequent distortion of the sound field were not present) is placed 1 metre in front of and on the axis of the loudspeaker unit. The microphone feeds into a suitable amplifier and the amplifier operates the readout in terms of relative sound pressure.

The test should be carried out either in the open air or in an anechoic chamber (a room with acoustically absorbent boundaries avoiding reverberation).

When the audio system contains more than one loudspeaker unit, the measurement is made on the axis of the unit handling the high frequencies. With a complex loudspeaker system, however, a suitable axis somewhere towards the centre of the front baffle should be chosen.

After having established a 1kHz reference, the sound pressure can be measured over the audio spectrum and at each test frequency a plot can be made to develop a pressure response curve as shown in Fig. 5.26.

### Directional Characteristics

To determine the directional characteristics of the loudspeaker of an audio system or of a loudspeaker system the setup in Fig. 5.25 can be used, but this time the sound pressure is measured as a function of the direction of the microphone relative to the loud-

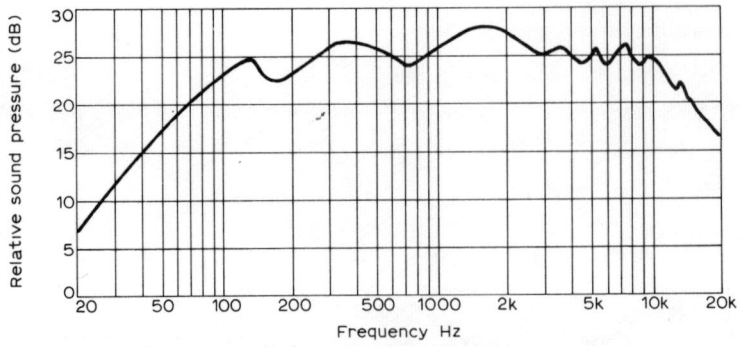

Fig. 5.26   Sound pressure curve measured with the setup in Fig. 5.25.

# System Tests

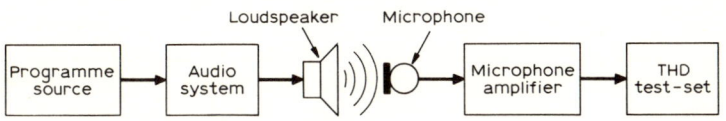

Fig. 5.27 Setup for measuring overall THD. The same scheme can be adopted for measuring parameters such as IMD and frequency response.

speaker. The angle from the axis is measured and plotted at each test, and a family of curves at various frequencies can be constructed on polar graph paper carrying $\frac{1}{16}$ in. co-ordinates.

## Tests for Overall Distortion, etc.

Fig. 5.27 shows a setup suitable for measuring the THD through a complete system, including the programme source and loudspeaker(s). The measurement procedure is as already detailed, but in this case with tone produced by the loudspeaker is picked by the microphone and fed, via a microphone amplifier, to the THD test-set.

Based on the same idea, tests can be made of IMD and frequency response. These can be very useful when properly performed since they give a true objective 'picture' of the overall performance of a system such as a tape recorder, record reproducer or, indeed, hi-fi system.

CHAPTER SIX

# AUDIO STANDARDS AND DEFINITIONS

NUMEROUS ORGANISATIONS throughout the world are involved in the vast subject of audio standards and their progressive development in every area of activity. Some standards have been completed and are in current application; some are being upgraded or evolved, while others at the time of writing consist merely as interface parameters in draft form.

As would be expected, audio standards can never remain static for very long. New ones have to be produced to satisfy contemporary techniques, while earlier ones need to be updated to match the increasing quality demands of audio reproduction. Owing to the vastness of the subject and its state of flux, it is impossible even to mention all the organisations concerned, let alone to list all the standards.

However, the four standards organisations of chief interest to the audio technician are: British Standards Institution (British Standards House, 2 Park Street, London W.1.), Institute of High Fidelity, Inc. (516 Fifth Avenue, New York, USA), Commission Electrotechnique Internationale (International Electrotechnical Commission) (Central Office of the IEC, 1 rue de Varembé, Geneva) and Deutscher Industrie Normenausschuss (DIN) of Germany.

BSI, IHF and IEC standards deal essentially with methods of measurement, while the various DIN standards deal both with methods of measurement and with the establishment of a basic minimum requirement for hi-fi equipment, this latter being known as DIN 45-500.

At the time of writing DIN 45-500 is considered ready for upgrading, but little has been published altering the original context, and since a great deal of domestic audio and hi-fi equipment is based on this standard, it is possibly one of the most important even though it cannot be regarded as a measure of top hi-fi performance.

## Audio Standards and Definitions

DIN 45-500 is, in fact, the only standard setting out to establish minimum requirements (or any other) for hi-fi equipment. However, it seems likely that a new standard to serve a similar purpose will ultimately emerge by way of the Audio Specification Co-ordinating Committee (ASCC), which is currently drafting specifications for domestic and industrial sound reproducing equipment.

The ASCC has already evolved interface parameters, and when these have been approved by mutual agreement between the constituent Associations of the ASCC they will be submitted to the BSI for consideration.

Thus, while there is a definite move afoot to establish a new set of audio standards (and let us hope that they will be significantly in advance of DIN 45-500 at this present state of the art), it is likely that this book will have been in print for some time before they are are made 'official' through the BSI. What is really wanted, of course, is a world-wide standard for the various parameters of hi-fi equipment or, at least, a uniform European one.

The various other standards dealing with measurement often differ both in definitions and measurement techniques which, again, is a pity. Nevertheless, there are a number of areas of reasonable conformity, and the various definitions involved will be referred to later.

## DIN 45-500

For the moment let us concentrate on DIN 45-500. This is covered in nine sections as follows:

1. General Requirements.
2. Minimum Requirements for F.M. Receivers.
3. Minimum Requirements for Record Players.
4. Minimum Requirements for Magnetic Tape Recorders.
5. Minimum Requirements for Microphones.
6. Minimum Requirements for Amplifiers.
7. Minimum Requirements for Loudspeakers.
8. Minimum Requirements for Combined Units.
9. Minimum Requirements for Magnetic Tapes.

Let us look briefly at the important ones of these, starting with No. 1.

### General Requirements

*Purpose Range and Intention.* This section sets out the scope of the standards, indicating clearly that they are the *basic minimum* requirements for equipment intended for home, but not professional, use.

*Climatic Conditions.* These are the conditions of measurement and give the ambient temperature as 15–35 deg. C, the relative humidity as 45–75% and the pressure as 860–1060mbar.

*Adaptability.* Here is emphasised the necessity for 'standard' connections, uniform tolerance ratings and compatibility of the various components of a system.

*Labelling.* The importance of correct and adequate function indications is stated, on the items of equipment themselves or, at least, in the operating instructions.

### F.M. Tuners

*Range of Reproduction.* This is given as 40Hz-12·5kHz with permissible deviations of ±3dB 40–50Hz, ±1·5dB 50Hz-6·3kHz and ±3dB 6·3–12·5kHz.

*Difference Between Stereo Channels.* Maximum permitted is 3dB from 250Hz to 6·3kHz.

*Harmonic Distortion.* Less than 0·2% at 1kHz modulation frequency and ±40kHz deviation, based on identical left and right channel signals with stereo tuners.

*Stereo Crosstalk (i.e. Channel Separation).* Better than 26dB from 250Hz to 6·3kHz and 15dB from 6·3 to 12·5kHz.

*Disturbance Ratios.* These are referred to the output voltage produced by a modulation signal of 1kHz and ±40kHz deviation, and include the next two items.

*S/N Ratio (Unweighted).* Better than 54dB over the range 40Hz to 15kHz, mono *and* stereo.

*Pilot Tone S/N Ratio (Unweighted).* Better than 20dB at 19kHz and 30dB at 38kHz, when measured selectively with the aerial input correctly loaded and with a deviation of ±67·5kHz.

*General Parameters.* These, which should be stated, include:

(i) Aerial input impedance.
(ii) Aerial voltage (presumably e.m.f.) at the recommended impedance.
(iii) Audio output impedance and permitted loading values.
(iv) Output voltage at ±40kHz deviation.

*Connectors (Inputs and Outputs).* These should be of the DIN standards, with connections as shown in Fig. 6.1 for tape recorder replay and record. Other DIN connections are given in the *Tuners and Amplifiers* book.

# Audio Standards and Definitions

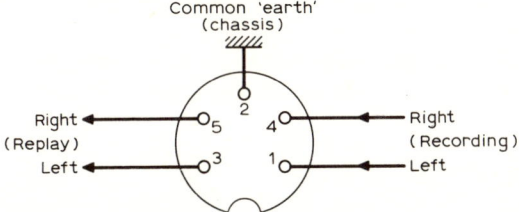

Fig. 6.1  DIN connections for recording and replay, left and right channels.

## Record Players

*Turntable Speed Variations.* These are given as $+1\cdot5\%$ and $-1\%$.

*Variations from Standard.* When established with the DIN 45-545 disc should not exceed $\pm 2\%$.

*Rumble.* Referred to 1kHz and 10cm/S velocity a value better than 35dB should be obtained using the DIN 45–544 test disc. The noise-voltage ratio is given as at least 55dB, using the same disc, frequency and velocity.

*Range of Reproduction.* At least 40Hz to 12·5kHz with permissible deviations of $\pm 5$dB between 40–63·5Hz, $\pm 2$dB between 63·5Hz–8kHz and $\pm 5$dB between 8–12·5kHz. A DIN test disc 45–541 is available for these measurements.

*Stereo Balance.* Better than 2dB at 1kHz, using the DIN 45–541 disc for measurement.

*IMD.* This is given as 1% based on the use of an 'audimeter', as described in DIN 45–507. A suitable test disc is the DIN 45–542.

*Channel Separation.* Better than 20dB at 1kHz and 15dB in the range 500Hz–6·3kHz. DIN 45-543 disc is suitable for this test.

*Tracking Weight of Cartridge.* Not to exceed 5 pond (approximately equivalent to 5 grams).

*Compliance of Cartridge.* At least 4 c.u.s in each axis (i.e. $4 \times 10^{-6}$cm/dyne).

*Effective Tip Mass of Cartridge.* This is given as 2 milligrams, but the standard notes the inter-relationships of other cartridge parameters in terms of tracking performance, etc. (see Chapter 4).

*Tip Dimensions (Cartridge).* 15µm $\pm$ 3µm when spherical and 6µm $\times$ 20µm when biradial.

*Vertical Tracking Angle of Cartridge.* Given as the 'standard' 15 degrees, with a tolerance of $\pm 5$ degrees, based on the use of the distortion test disc DIN 45–542 (also see Chapter 4).

*Output Voltage.* Based on a velocity of 10cm/S (1kHz) high output players should yield 500mV to 1·5V into a 470kΩ load, while low output players (based on a magnetic cartridge) should yield 8 to 20mV into a 47kΩ load. Cartridge type to be stated.

**Tape Recorders**

Tape recorder characteristics are described in a separate standard, called DIN 45-511, but DIN 45-500 gives in addition the following requirements.

*Speed Variation (Above 30 sec.).* $\pm 1\%$ maximum.

*Momentary Speed Variations (Wow and Flutter).* $\pm 0.2\%$ maximum.

*Reproduction Range.* 40Hz–12·5kHz, within the limits shown in Fig. 6.2.

*Distortion.* 5% when measured at 333Hz and full modulation.

*S/N Ratio (Tape Stationary).* At least 50dB referred to maximum output.

*S/N Ratio (Overall).* At least 45dB referred to full tape modulation and output.

*Stereo Separation.* At least 25dB at 1kHz.

*Track Crosstalk.* At least 60dB at 1kHz.

*Erasure Efficiency.* At least 60dB.

**Microphones**

*Reproduction Range.* This is given as 50Hz–12·5kHz, with the optimum response lying within the limits defined in Fig. 6.3. Below 250Hz the broken-line gives the limits for non-directional microphones and the full-line for directional microphones. The response variations within an octave should not exceed 3dB. Tolerances are: $\pm 4$dB 50–250Hz, $\pm 3$dB 250Hz–8kHz and $\pm 4$dB 8–12·5kHz.

*Directional Characteristics.* The standard states that these should not be influenced unduly by frequency, and that the frequency curves for all angles of response, relative to 0 degree, should exhibit only small variations.

*Non-Directional Microphones.* Placed in free field sound over the range 6–9kHz the output should not change by more than 12dB at any angle between 0–90 degrees.

*Directional Microphones.* The focusing degree should be greater than 2 over the frequency range 250Hz–8kHz, and in the same frequency range the response amplitude for any angle relative to 0 degree should not differ by more than $\pm 4$dB.

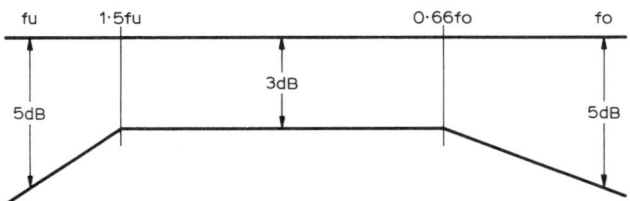

Fig. 6.2  Frequency response limits for tape recorders (DIN 45–500).

*Distortion.* This should not exceed 1% within the frequency range 350Hz–8kHz at a sound pressure of 100μbar (114dB).

*Channel Balance.* Should be better than 3dB in the frequency range 250Hz–8kHz.

**Amplifiers**

This section takes into account preamplifiers, power amplifiers and integrated amplifiers. The two channels of stereo amplifiers should be considered under equal signal conditions. When necessary, the controls should be adjusted for the correct or specified input impedance, but the volume control should normally be in the maximum position, the tone controls 'flat' and the filters switched out, unless required otherwise by the tests.

*Range of Reproduction.* 40Hz–16kHz with permissible variations of $\pm 1\cdot 5$dB at 'flat' (i.e. not equalised) inputs and $\pm 2$dB at equalised inputs, when the output is 6dB below maximum.

*Channel Balance.* Normally better than 3dB, but better than 6dB permissible when the equipment features a balance control of better than 8dB range.

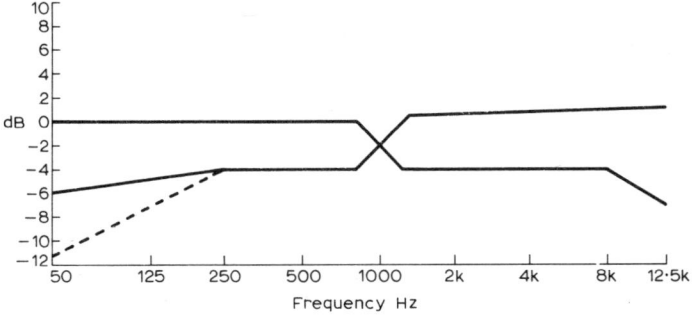

Fig. 6.3  Limits for microphones (DIN 45–500).

*Harmonic Distortion.* (i) Preamplifiers: 1% maximum at full output over the range 40Hz–4kHz.

(ii) Power Amplifiers and Integrated Amplifiers: 1% maximum from full power down to 20dB below full power over the frequency range 40Hz–12·5kHz.

*Intermodulation Distortion.* 3% maximum when measured with test frequencies of 250Hz and 8kHz in the amplitude ratio of 4:1 (see Chapters 1 and 2).

*Stereo Separation.* At least 50dB at 1kHz and at least 30dB over the frequency range 250–10kHz.

*Crosstalk Between Inputs.* At least −50dB at 1kHz and at least −40dB over the frequency range 250Hz–10kHz.

*S/N Ratio.* (i) Preamplifiers: At least 50dB referred to the nominal input voltage at 1kHz.

(ii) Power Amplifiers and Integrated Amplifiers: At least 50dB up to 20W rating, referred to 100mW output, and proportionally higher (based on the power increase in decibels) when the rating exceeds 20W.

*Power Output.* At least 10W mono and $2 \times 6$W stereo. The amplifier must be capable of sustaining a 1kHz sinewave signal at the rated power for at least 10 minutes.

*Nominal Input Voltages and Overload.* The input required to produce the rated power is the nominal input (at 1kHz), and in equipment where the input signal passes through a volume control the non-linear distortion should not exceed 1dB of that specified when the input is increased by 12dB.

*Input Impedances.*

Non-equalised: 470kΩ or more up to 0·5V.

Equalised: 47kΩ or more up to at least 5mV at 1kHz.

Smaller input impedances are permissible provided the influence on the response is taken into account.

*Outputs.* (i) Preamplifiers: 1V or more at 47kΩ or more.

(ii) Tape Recording: 0·1–2mV for every 1kHz of load resistance from 1kΩ to 50kΩ. 40mV at 47kΩ is common.

(iii) Loudspeakers: Preferred impedances are 4, and 16Ω. Other intermediate values are also stated.

**Loudspeakers**

Loudspeaker tests are made under free field conditions and away from reflecting materials such that the sound waves can travel parallel to the ground and radiate freely upwards.

*Range of Reproduction.* This is given as 50Hz–12·5kHz under specified conditions of measurement when the measuring instrument is placed 1 or 3 metres from the loudspeaker. The response characteristic should resemble a horizontal line between 100Hz and 4kHz, with limits determined through points 8dB above the centre line. The curves in Fig. 6.4 show the tolerances, beyond which the measured curve should not deviate at any point.

Stereo loudspeaker pairs should not differ from each other by more than 3dB over the range 250Hz–8kHz.

*Sound Pressure.* The loudspeaker should be capable of producing a sound pressure of 96dB (12μb) at a distance of 1m or about 86dB (4μb) at a distance of 3m.

*Distortion.* When the loudspeaker is correctly terminated, this should not exceed 3% 250Hz–1kHz, 3% proportionally to 1% 1–2kHz and 1% over 2kHz, tests being made up to 5kHz.

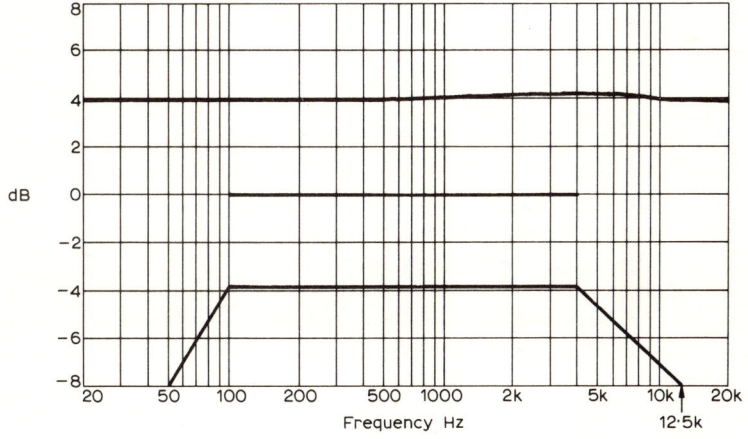

Fig. 6.4 Loudspeaker tolerance curves (DIN 45–500).

*Operating Capacity.* The rated power capacity should be accommodated with the signal applied for one minute and removed for two minutes, over a period of 300 hours. Due allowance must be made for the thermal characteristics of sealed enclosures.

*Loading Limits.* This is defined as minimal vibration and distortion resulting from sinewave input from 250Hz down to the lowest limit applied at short intervals not exceeding two seconds.

*Impedance.* This should not exceed the nominal by more than 20% at any frequency within the reproduction range.

## SETS AND SYSTEMS

The tests are made in this section to determine the performance of a *complete system*, and with stereo systems both channels are supplied with the common signal. Frequency controls are adjusted to 'flat' so that the specified frequency tolerances are satisfied and the volume control is set for maximum amplification.

### Amplification Requirements

*Reproduction Range.* This is given as 40Hz–16kHz with permitted deviations from 1kHz of $\pm 1\cdot 5$dB on non-equalised inputs and $\pm 2$dB on equalised inputs, when the response is measured at 6dB below full output. The test output is taken from the loudspeaker terminals at the specified load impedance.

*Channel Balance.* Better than 3dB or 6dB when the system incorporates a balance control, these tolerances to be maintained from maximum setting of the volume control down to $-40$dB. The tests are made at $-6$dB full output.

*Harmonic Distortion.* 1% maximum over the power bandwidth range of 40Hz–12·5kHz and from full output to $-20$dB.

*Intermodulation Distortion.* 3% maximum using test frequencies of 250Hz and 8kHz in the amplitude ratio of 4:1.

*Channel Separation.* At least 40dB at 1kHz.

*S/N Ratio (Unweighted).* At least 50dB referred to 100mW with ratings up to 20W and proportionally greater for rating above 20W.

*Power Output.* At least 10W mono and 2 × 6W stereo. 1kHz sinewave power should be sustained for at least 10 minutes.

### Record Players with Amplifiers

The mechanical features and parameters of these are akin to those detailed under 'Record Players'.

### Electrical Characteristics

*Reproduction Range.* 40Hz–12·5kHz with permissible deviations of $\pm 6\cdot 5$dB 40–63·5Hz, $\pm 3\cdot 5$dB 63·5Hz–8kHz and $\pm 6\cdot 5$dB 8–12·5kHz, as measured with the volume control adjusted for $-6$dB of full power.

*Channel Balance.* Better than 5dB except where a balance control

## Audio Standards and Definitions

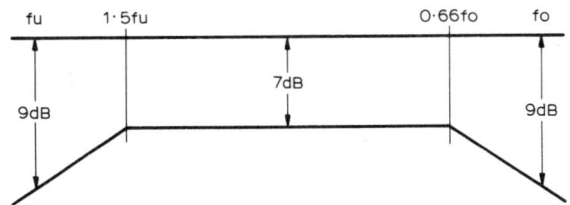

Fig. 6.5 Frequency response limits for tape systems (DIN 45–500).

is incorporated providing a range of 8dB at least, when the ratio can be 8dB.

*Harmonic Distortion.* As under this parameter in 'Amplification Requirements'.

*Intermodulation Distortion.* As under 'Record Players'.

*Channel Separation.* Not less than 19dB at 1kHz and 14 dB over the frequency range 500Hz–6·3kHz.

### Pickup Parameters

As detailed under the appropriate parameters for 'Record Players'.

### Mechanical Parameters

As detailed under the appropriate parameters for 'Record Players'.

### Tape Recorders with Amplifiers

The mechanical characteristics of the tape recorders should satisfy those detailed under 'Tape Recorders'.

### Electrical Characteristics

*Reproduction Range.* 40Hz to 12·5kHz when measured −20dB full power. With the volume control at maximum on replay, the response should fall within the tolerances shown in Fig. 6.5.

*Distortion.* 5% at 333Hz at full volume.

*Stereo Separation.* At least 24dB at 1kHz and 21dB between 250Hz and 10kHz.

*S/N Ratio (Unweighted).* At least 41dB.

*Output Power.* At least 10W mono and 2 × 6W stereo. The sinewave power must be sustained for at least 10 minutes.

*Erasure Efficiency.* Not less than 60dB.

### Tape Replay Systems with Amplifiers (No Recording)

The mechanical parameters are as already detailed, while the electrical characteristics conform essentially to those of amplifiers.

### F.M. Tuners with Amplifiers

This class of equipment includes the popular tuner-amplifier. Measurements are made, unless stated otherwise, with an aerial input power of $4 \cdot 16 \times 10^{-9}$W, corresponding to 1mV into 240Ω. The output is $-6$dB full volume, except when distortion is measured.

*Reproduction Range.* 40Hz–12·5kHz, with deviations referred to 1kHz (0dB) of $\pm 4 \cdot 5$dB 40–50Hz, $\pm 3$dB 50Hz–6·3kHz and $\pm 4 \cdot 5$dB 6·3–12·5kHz.

*Channel Balance.* Better than 6dB over the frequency range 250Hz–6·3kHz or 9dB when the equipment incorporates a balance control of 8dB range.

*Distortion (Harmonic).* 2·5% or less at 1kH modulation and $\pm 40$kHz deviation over a power bandwidth of 40Hz–12·5kHz.

*Stereo Separation.* 24dB or more at 1kHz, 18dB or more over 250Hz–6·3kHz and 14dB or more over 6·3–10kHz.

*Disturbance Ratios.* The following ratios are referred to 100mW (2 × 50mW stereo equipment) with the volume control adjusted to give the operating level at 1kHz and $\pm 40$kHz deviation for equipment up to 20W power rating. With more powerful equipment the dB ratio is increased proportionally.

*S/N Ratio (Unweighted).* 41dB or more over the frequency range 40Hz–15kHz.

*Noise Voltage Ratio.* 50dB or more over the same frequency range.

*Pilot Tone S/N Ratio (Rejection).* With input signal deviated to $\pm 67 \cdot 5$kHz and measured selectivity, the ratio at 19kHz should be 19dB or more and at 38kHz 29dB or more.

*Output Power.* At least 10W mono and 2 × 6W stereo. Power should be maintained with 1kHz sinewave for at least 10 minutes.

This, then, concludes the main features of DIN 45-500. In some respects it will be seen that the requirements fall significantly short of advanced hi-fi reproduction; but it is stressed that the standards are for *minimum* requirements only. At the time of writing revised DIN specifications were not available but these original ones will be of value for some years yet to come as vast items of equipment have been based on them over a number of years.

It is noteworthy that even those items of equipment whose parameters are based on DIN 45–500 commonly exhibit a measured performance in advance of the minimum requirements.

## Audio Standards and Definitions

### ASCC Standards

At the time of writing the hi-fi parameters drafted by the ASCC (for BSI consideration) were not available for publication.

### British Standards

As already mentioned, the majority of the various other standards deal essentially with test methods. The two main BSI documents are BS3860:1965 entitled 'Methods for Measuring and Expressing the Performance of Audio-Frequency Amplifiers for Domestic, Public Address and Similar Applications' and BS4054:1966 entitled 'Methods for Measuring and Expressing the Performance of Radio Receivers for AM and FM Sound Broadcast Transmissions'. There is also a complementary document dealing with loudspeakers, BS2498, entitled 'Recommendations for Ascertaining and Expressing the Performance of Loudspeakers by Objective Measurements'.

### IHF Standards

The two main IHF documents are IHF-A-201:1966 entitled 'IHF Standard Methods of Measurement for Audio Amplifiers' and IHFM-T-100:1958 entitled 'IHFM Standard Methods of Measurement for Tuners'.

### IEC Standards

The Central Office of the IEC have issued standards dealing with 'Recommended Methods of Measurement on Radio Receivers' and 'Audio Frequency Measurements'.

Sadly, differences do exist between the test methods recommended, and some of these are highlighted in the foregoing chapters of this book. The IHF methods seemingly differ mostly from the others, while those of IEC are sometimes compatible with BSI.

### Terms and Definitions

There are also differences between the terms and definitions adopted by the various authorities, with the maximum difference between those of IHF and BSI. The important ones of these have also been revealed in the foregoing chapters.

Thus, when comparing the parameters of a specification based, say, on the IHF test methods with those of a specification based on IEC or BSI account must be taken of the different methods of test.

An important difference is the modulation deviation (depth) between IHF and BSI, the former often being 100% and the latter 30%. This can make a difference of almost 10dB in the readout of some parameters.

The IHF 'music power' parameter also differs significantly from the BSI, IEC and DIN continuous power (based on steady-state signal) parameter. R.f. test frequencies and signal powers (at the input of an f.m. tuner, for example) differ between BSI and IHF.

There are numerous other examples of differences in detail if not in principle. One cannot be dogmatic and state conclusively that one method of testing is better than another; but it seems such a pity that there is not a world-wide standards authority. With the different authorities giving their recommendations at random this seems to be some way off at the moment.

Moreover, none of the standards has yet caught up with the state of the art, and many parameters needing measurement for objective appraisal do not exist. Over recent years I have found it necessary to evolve my own methods of measuring these and gladly present them in this book for the assistance of fellow engineers and technicians.

# INDEX

Acoustic feedback test, 155
Acoustical characteristics of audio system, 163, 164
Aerial input signal levels, 85
A.f.c. pull-in range, 97
A.m.: artificial aerials, 103
   frequency response, 107
   sensitivity, 166
   signal-to-noise ratio, 106
   suppression, 77, 92
   test frequencies (BSI), 102
   test modulation, 106
   test signal voltages, 103
   volume sensitivity (IHF), 106
Amplifier: power capacity, 35
   rated output, 36
   tests, 33–71
Amplifiers: Class A, 6
   Class B, 5, 6, 17
   overshoot, 10
   ringing, 10
   wide power bandwidth, 10
Artificial aerials, 84, 85, 104, 105, 138
ASCC Standards, 177
Attenuation plotting, 79
Audio: generators, 21
   millivoltmeters, 75
   oscillators, 57, 63, 65, 72
   output and limiting, 73
   signal filtering, 76
   standard and definitions, 166–178
   voltmeters, 15, 16, 57
   waveforms, 69
Automatic frequency control, 96
Automatic lock on wave analysers, 50

Band II, 72
Bandpass filter, 82
Bearing friction, tests for, 125, 126
Beat interference, tuners, 96
British standards, 177

Capture effect, 88
Capture ratio (IHF), 87
Cartridge hum induction test, 136
Cartridge minimum tracking weight, 111
Cartridge trackability, 108, 109
Channel balance tests, 147, 148
Class A amplifier, 39
Class A amplifier, causes of distortion in, 52, 53
Class B amplifiers, 46
Class B amplifiers, causes of distortion in, 52, 53
Class B amplifiers, crossover distortion, 20
Comparative tests, 2
Correx force gauge, 127
Coupling to ferrite rod aerials, 103
Crossover distortion, 54
Crossover network, 10
Crosstalk (between inputs), 24

Damping, aspects of, 63
Damping factor, 14, 38, 60–62
Decoder output, 81
De-emphasis, 78
Detector non-linearity, 77
Deviation (f.m.), 75, 81
Deviation frequency, 83

DIN 45–500, 167–176
Disc: playing equipment tests, 108–136
  ripple, 110
  rumble, 110
  speed stability test, 135
  tone-burst and squarewave tests, 122
Distortion factor, 47
Distortion factor meter, 4, 17, 18, 44
Distortion measuring set, 49
Distortion test sets, 22, 63
Drive preset, 39
Dummy aerials, 105
Dummy load, 13, 38
Dummy load resistors, 20
Dynamic output, 14

Eagle K1400 multirange testmeter, 24
Effective mass tests, 128, 129
Electromagnetic control, 62
European pre-emphasis, 78

Ferrite rod aerial, 105
Filters, 20, 60, 144, 145
F.m.: generators, 72, 77
  muting levels, 28
  stereo generator, 97
  stereo tests, 97
  stereo signal/noise ratio, 101
  test frequencies (BSI), 74, 75
  tuner tests, 72, 73
Frequency response, 20, 78
Frequency response measurement, 57

Generator matching, 84
Grundig digital voltmeter, DV33A, 26

Harmonic distortion, 73, 131
Harmonic distortion measurements, 17
Heathkit audio analyser, IM-48, 31
Heathkit f.m. stereo generator, 100
Heathkit sine-square audio generator, Model IG-18, 27
High-frequency distortion, 55
High-pass filter, 12, 18
High-wattage variable resistor, 15
Hum and noise, 65–67
Hum-loop conditions, 59, 66

IEC standards, 177
I.f. limitations in amplifiers, 36
IHF: procedures, 140, 141
  standards, 177
Image rejection ratio, 94

IMD test, 161
Impulsive interference, 145, 146
Input amplifier overload, 22
Input sensitivity voltage, 21
Input overload, aspects of, 64
Intermodulation components, 95
Intermodulation distortion, 18–20, 50, 51
Instability, 37
Institute of High Fidelity (IHF), 13, 14
Intermediate frequency rejection ratio, 94
Instruments: for amplifier tests, 13–25
  for disc and tape tests, 30, 31
  for tuner tests, 25–29

J. E. Sugden audio instruments, 19, 20 30, 59
Lateral tracking error, 130
Limiting, 75
Low-frequency distortion, 55
Low-noise input, 47, 48
Low-pass filter, 51
Loudspeaker impedance test, 156
Loudspeaker testing, 2

Magnetic cartridge loading, 116
Magnetic cartridge response, 119
Maximum modulation, 75
Measuring: amplifier overload voltage, 63
  arm mass, 129, 130
  audio output, 73
  capture ratio, 88
  f.m. tuner frequency response, 78
  f.m. tuner THD, 76
  frequency response, 117
  hum induction of pickup, 135
  IMD, 51
  lateral friction of a pickup arm, 125
  limiting f.m. tuner, 73
  pickup output voltage, 123
  pickup tracking, 109
  power output, 35
  selectivity, 89, 91
  signal-to-noise ratio, tuner, 82
  stereo identicalityfactor (BSI), 149
  THD, 45
  transient performance, 123
  rumble, turntable unit, 133
  usable sensitivity (IHF), 86
Millivoltmeter, 16
Mistracking, 111

# Index

Modulation depth, 83
Modulation frequency, 73, 75
Mono tuner, 80

Negative feedback, 60, 61
Noise generator, 22
Noise input voltage, 66
Noise-limited sensitivity, 85
Noise measurements, 8
Non-integrated amplifiers, 34, 35, 64
Non-linearity, 77

Oscillator distortion, 77, 78
Oscillograms:
  assymmetrical clipping, 37
  distortion, 7, 44
  mistracking, 111
  overshoot, 37
  pulsed-tone, 123, 124
  ringing, 8, 11
  rise/fall time, 68
  squarewave, 9, 70, 71
Oscilloscope (Heathkit OS-2), 29
  for waveform analysis, 6–11, 47
Output signal voltage, 21
Output impedance, 15
Overall rejection, 81
Overload, 64
Overload performance, 65
Overshoot, 71

Parameters, f.m. tuning, 77
Peak power, 38
Phase-shift test, 116
Pickup: arm tests, 125
  frequency response tests, 116
  cartridge tests, 108
  distortion tests, 132
  output voltage, 123
Piezo cartridge, 116
  frequency response, 119
Power output tests, 13
Power bandwidth tests, 14, 55–57
Power supply filters, 37
Preamplifier signal levels, 21, 22

Quasi-complementary power amplifier, 40

Radford low distortion audio oscillator, 21
  distortion measuring set, 16

Radio breakthrough susceptibility test, 142–144
Recording signal levels, 21
Residual hum, 47
Response troubles, 60
R.f. signal levels, 72, 82
Rheostat for power load, 15
RIAA preamplifier, 64, 67
  recording characteristics, 59, 113, 119, 120
Rise time, 69
Rogers audio millivoltmeter, 23
  distortion factor meter, 18
  low distortion oscillator, 23
Rumble tests, 132, 133

Standard signal input level, 91
Selectivity, 89
  BSI, 89, 90
  IHF, 92
Sensitivity voltage, 6
  usable (IHF), 86
Series regulator, 43
  resistor, 81
Side-thrust correction (pickups), 113
  force tests, 126, 127
Signal/noise ratio, 22, 65, 82
Sinewave generator, 13
Solidstate equipment, 67
Source resistance, 14
Squarewave tests, 67–69
Stereo amplifiers, 13, 36, 37
  decoders, 77
  decoder performance, 97
  disc separation limits, 121
  f.m. tuners, 97
  instruments, 100, 101
  subchannel rejection, 81
  switching, 28, 102
  tuner crosstalk performance, 98
Stylus tip impedance, 112
Subchannel attenuation, 81
Switched attenuator, 58, 79
Symmetrical clipping, 40
System tests, 137–165

Tape recorder tests for:
  frequency response, 162, 163
  distortion, 160
  signal/noise ratio, 161
  wow and flutter, 163
Test discs, 30, 110, 117
Test instruments, 1, 32

# Index

Test laboratory, 23
Test tapes, 31, 126
Test setups for:
  amplifier overload voltage, 63
  capture ratio (IHF), 27, 87, 88
  channel balance, 147, 149
  compliance (dynamic), 2, 115
  compliance (static), 115
  crosstalk (system), 152
  damping factor, 61
  disc frequency response, 25, 57, 58, 78, 117, 158, 163
  harmonic distortion, 17, 28, 45, 76
  hum induction (pickup), 135
  hum level, 155
  image response, 27
  impedance (of loudspeaker), 156
    impedance (of system input), 157
  intermodulation distortion, 51, 165
  limiting, 101, 141
  overall channel balance, 148
  phase tests, 150, 151
  pickup output, 123
  power capacity, 35, 42
  radio breakthrough susceptibility (of amplifier), 143
  recording (tape):
    head bias, 161
    head current, 161
  rumble (record playing unit), 133
  selectivity, 28, 89–92
  sensitivity (general), 25
    (f.m. tuner IHF), 86, 141
  signal/noise ratio, 25, 64, 82, 139
  spurious response (f.m. tuners), 27
  squarewave tests, 67
  tracking (pickup), 109
Third harmonic distortion, 67, 76, 95
Time-constant, 78, 79
Time period, 68

Tip mass, 115
Tone control tests, 59
Total harmonic distortion, 3, 4, 13–18, 22, 36, 65, 72, 87, 165
  measuring for, 44–50
  meters for, 15
  readout, 46, 47
  residue waveforms, 6, 7
Tracking weight and counterbalance tests (pickup), 130
Tracking test records, 114
Transient tests, 9, 10, 25
Transistor amplifiers, 14, 60
  temperature, 54
Treble filter, 71
Tuner-amplifier tests, 139
Tuner crosstalk preset, 99
Tuner input levels, 93
Tuner test tone, 98, 99
Tuning indicator, 77

Usable sensitivity (IHF), 86

Vertical tracking error (pickups), 131
V.h.f. carrier frequencies, 73

Wave analyser, 17, 18
Wave analysis, 49, 50
Waveforms, 69–71
Waveform clipping, 63
Weighting curve and circuit, 8, 9
Wow and flutter tests, 134

X axis, 74

Y bandwidth of oscilloscope, 32
Y amplifier, 68, 69

Zero reference, 58